Android 应用程序开发教程
——理论、实验与在线金课（第 2 版）

罗文龙 编著

电子工业出版社
Publishing House of Electronics Industry
北京·BEIJING

内 容 简 介

本书以 Google 推出的 Android IDE - Android Studio 和 Android 模拟器 Genymotion 作为开发环境进行编写，全面介绍了 Android 应用开发的相关知识，内容覆盖了 Android 系统与开发环境，Android UI 设计，基本程序单元 Activity，Android 应用核心 Intent 和 Filters，Android 事件处理，Android 服务，Android 广播接收器，ContentProvider 实现数据共享，图形、图片与多媒体，Android 网络编程基础，Android 数据存储，GPS 应用开发，以及对应的实验项目等。

本书不局限于介绍 Android 编程的各种理论知识，而是以"实例驱动"的方式来讲解。全书共 50 多个实例，这些实例能够帮助读者更好地理解 Android 的各种知识在实际开发中的应用。第 2 版中增加了 7 个 Android 实验项目，其内容与理论部分相匹配，为读者理解知识点提供实践支撑。与本书配套的所有实例和实验项目都可以登录华信教育资源网（www.hxedu.com.cn）注册后免费下载。同时，读者可以访问"学银在线"（www.xueyinonline.com）观看"智能终端应用程序开发"在线金课。该在线金课包含在线资料、在线作业、在线考试、在线讨论、在线直播、公告通知、在线活动和课程统计等模块。

本书可作为高等院校计算机科学与技术、软件工程、信息管理、电子商务等相关专业的本科生和研究生教材，也可供从事移动开发的工作者学习参考。

未经许可，不得以任何方式复制或抄袭本书之部分或全部内容。
版权所有，侵权必究。

图书在版编目 (CIP) 数据

Android 应用程序开发教程：理论、实验与在线金课 / 罗文龙编著. —2 版. —北京：电子工业出版社，2021.8

ISBN 978-7-121-41804-4

Ⅰ. ①A… Ⅱ. ①罗… Ⅲ. ①移动终端—应用程序—程序设计—高等学校—教材 Ⅳ. ①TN929.53

中国版本图书馆 CIP 数据核字（2021）第 165643 号

责任编辑：秦淑灵　　　　　　特约编辑：田学清
印　　刷：北京虎彩文化传播有限公司
装　　订：北京虎彩文化传播有限公司
出版发行：电子工业出版社
　　　　　北京市海淀区万寿路 173 信箱　　邮编：100036
开　　本：787×1 092　1/16　印张：20.5　字数：538 千字
版　　次：2016 年 8 月第 1 版
　　　　　2021 年 8 月第 2 版
印　　次：2025 年 2 月第 5 次印刷
定　　价：59.00 元

凡所购买电子工业出版社图书有缺损问题，请向购买书店调换。若书店售缺，请与本社发行部联系，联系及邮购电话：（010）88254888，88258888。

质量投诉请发邮件至 zlts@phei.com.cn，盗版侵权举报请发邮件至 dbqq@phei.com.cn。
本书咨询联系方式：qinshl@phei.com.cn。

前　言

随着移动互联网的快速发展，无论是个人的工作还是生活，都受到其极大的影响。移动互联网的时代已经到来，移动互联网已成为全世界商业和科技创新发展的加速器，成为当今时代最大的机遇和挑战。

Android 就是一个开放式的移动互联网操作系统。目前，Android 已经成为移动互联网的"宠儿"，是应用最广泛的移动互联网平台（根据 2020 年的数据统计，Android 的市场占有率为 74.6%，iOS 的市场占有率为 24.82%）。因此，手机软件在当今的 IT 行业中具有举足轻重的地位。从招聘市场的情况来看，对 Android 软件人才的需求也越来越大。

在 2013 年 Google I/O 大会上，Google 正式推出了官方 Android 软件集成开发工具 Android Studio，并在 2015 年宣布停止对 Android Eclipse 的支持。以前，很多书籍都是以 Android Eclipse 为开发环境编写的，但以后基于 Android Studio 的 IDE 开发是大势所趋，所以本书采用 Android Studio 作为实例开发平台进行讲解。

本书注重讲解手机应用开发的新技术和新应用，突出先进性、系统性、实用性和可操作性，能够使读者在较短的时间内进行 Android 开发环境的搭建，深刻理解 Android 平台体系结构，熟练使用 Android 基本组件、Android 存储操作、多媒体开发、网络应用程序开发等技术，从而具备基本的算法设计能力、一定的系统设计和模块设计能力、一定的需求分析能力和软件代码编写能力。本书具有以下特色。

1．全新的开发环境

本书以 Google 官方 Android IDE-Android Studio V4.1.2 为开发环境对实例进行开发和讲解，让读者更快地了解 Android Studio 的界面操作。同时，本书引入了当前应用广泛的 Android 模拟器软件 Genymotion，并对 Genymotion 的安装、使用进行了详细介绍，让开发者摆脱 Android 模拟器运行缓慢、耗费内存的缺点，使 Android 开发更加得心应手。

2．由浅入深，循序渐进

本书以高等院校本科生为对象，使学生从了解 Android 和搭建 Android 开发环境开始，到学习 Android 开发的基础技术，再到学习 Android 开发的高级内容，最后学习如何开发一个完整项目。本书讲解步骤详尽、版式新颖，并在操作的内容图片上进行了标注，让读者在阅读时一目了然，从而快速掌握本书内容。

3．知识全面，覆盖面广

本书全面介绍了 Android 的相关知识，内容包括 Android 系统与开发环境，Android UI 设计，基本程序单元 Activity，Android 应用核心 Intent 和 Filters，Android 事件处理，Android 服务，Android 广播接收器，ContentProvider 实现数据共享，图形、图片与多媒体，Android 网络编程基础，Android 数据存储，GPS 应用开发，以及对应的实验项目等。

4．实例典型，轻松易学

通过实例进行学习是最好的学习方式，本书通过"实例驱动"的方式来讲解理论知识。

全书共 50 多个实例，这些实例能够帮助读者更好地理解 Android 的各种知识在实际开发中的应用。本书包含 7 个实验项目，综合运用前面所介绍的知识点，可以更好地帮助读者通过实践练习理解理论概念。

5．改革创新性强

本书采用"滚动在线课程视频、在线试题库、在线案例源代码分享、在线讨论小组和共享提升资源库"五位一体新平台教材编写模式。

编著者在编写本书时，将适应社会发展需要的新平台、新媒体、新应用等教学改革手段应用在本书中。2017 年编著者与北京世纪超星信息技术发展有限责任公司合作，共同开发并推出本书对应的线上"金课"。2018 年—2021 年在"学银在线"平台提供 6 期线上滚动课程"智能终端应用程序开发"，累计访问量为 300 957 人次，提供 42 节视频课、500 道在线试题、600MB 在线案例源代码、15 个在线讨论小组、1000 余个其他共享资源。上述在线资源在新冠肺炎疫情期间为各大高校的新常态教学方式提供了重要支持。

本书可作为高等院校计算机科学与技术、软件工程、信息管理、电子商务等相关专业的本科生和研究生教材，也可供从事移动开发的工作者学习参考。与本书配套的所有实例和实验项目都可以登录华信教育资源网（www.hxedu.com.cn）注册后免费下载。

特别感谢蹇洁教授对本书编写工作提供的指导和大力支持。同时感谢本书的编辑，没有她的策划、指导、无私帮助和辛勤工作，就不会有本书的出版。

本书仅以 Android Studio V4.1.2 + API 18/23 + Genymotion 为开发环境进行讲解，书中所论并不完美，难免存在错误和疏漏之处，恳请读者批评指正。编著者 E-mail：luowl@cqupt.edu.cn。

编著者

目　录

第 1 篇　Android 基础篇

第 1 章　Android 系统与开发环境 2
- 1.1　什么是 Android 2
  - 1.1.1　Android 平台的架构 2
  - 1.1.2　Android 系统的功能 5
  - 1.1.3　Android 系统分支 6
  - 1.1.4　Android 平台五大优势特色 7
- 1.2　Android Studio 入门 8
  - 1.2.1　Android Studio 的特点 8
  - 1.2.2　Android Studio 系统要求 8
  - 1.2.3　Android Studio 和 Eclipse ADT 的比较 9
- 1.3　Android 开发环境的搭建 9
  - 1.3.1　JDK 的下载 9
  - 1.3.2　JDK 的安装 11
  - 1.3.3　JDK 的环境变量配置 12
  - 1.3.4　在 Windows 环境下安装 Android Studio 15
  - 1.3.5　下载、安装和配置 Android SDK 17
- 1.4　Android 自带模拟器 AVD 19
  - 1.4.1　什么是 AVD 19
  - 1.4.2　通过 AVD Manager 创建 AVD 19
- 1.5　Android 模拟器 Genymotion 22
  - 1.5.1　什么是 Genymotion 22
  - 1.5.2　Genymotion 的特性 22
  - 1.5.3　Genymotion 运行环境要求 23
  - 1.5.4　注册和下载 Genymotion 模拟器 23
  - 1.5.5　安装 Genymotion 模拟器 25
  - 1.5.6　使用 Genymotion 模拟器 27
  - 1.5.7　给 Android Studio 安装 Genymotion 插件 29
- 1.6　新建一个 HelloWorld 项目 29

第 2 章　Android UI 设计 33
- 2.1　Android UI 布局 33
- 2.2　ListView（列表视图）... 43
  - 2.2.1　ListView（列表视图）的常用属性 43

2.2.2　ArrayAdapter（数组适配器）...... 44
2.3　GridView（网格视图）...... 46
2.4　Android UI 控件...... 53
　　2.4.1　TextView（文本框）...... 53
　　2.4.2　EditText（编辑框）...... 55
　　2.4.3　AutoCompleteTextView（自动填充文本框）...... 58
　　2.4.4　Button（普通按钮）...... 61
　　2.4.5　ImageButton（图片按钮）...... 63
　　2.4.6　CheckBox（复选框）...... 66
　　2.4.7　ToggleButton（开关按钮）...... 70
　　2.4.8　RadioButton（单选按钮）与 RadioGroup（按钮组）...... 73
　　2.4.9　使用 ProgressDialog（进度对话框）类创建 ProgressBar（进度条）...... 77
　　2.4.10　Spinner（列表选择框）...... 80
　　2.4.11　TimePicker（时间拾取器）...... 83
　　2.4.12　DatePicker（日期拾取器）与 DatePickerDialog（日期拾取器对话框）...... 87

第 3 章　基本程序单元 Activity 93

第 4 章　Android 应用核心 Intent 和 Filters 98
4.1　Intent 对象的各属性...... 98
　　4.1.1　Component（组件）...... 99
　　4.1.2　Action（动作）...... 99
　　4.1.3　Category（类别）...... 100
　　4.1.4　Data（数据）和 Type（类型）...... 101
　　4.1.5　Extra（额外）...... 102
　　4.1.6　Flag（标记）...... 102
4.2　Intent 的类型...... 103
　　4.2.1　显式 Intent...... 103
　　4.2.2　隐式 Intent...... 103
4.3　Intent Filters（意图过滤器）...... 107

第 5 章　Android 事件处理 115
5.1　Android 事件处理概述...... 115
5.2　事件监听器的注册方法...... 115

第 6 章　Android 服务 125
6.1　Service 的分类...... 125
6.2　Service 的生命周期...... 125
6.3　Service 生命周期中的回调方法...... 126

第 7 章　Android 广播接收器 131
7.1　创建 Broadcast Receiver...... 131
7.2　注册 Broadcast Receiver...... 131

| | 7.2.1 | 接收用户自定义 Broadcast Intent 消息 .... 132 |
| | 7.2.2 | 接收系统广播消息 .... 132 |

第 8 章 ContentProvider 实现数据共享 140
8.1 ContentProvider 概述 140
8.2 URI 简介 141
8.3 创建 ContentProvider 141

第 9 章 图形、图片与多媒体 152
9.1 基础绘图 152
9.1.1 常用绘图类 152
9.1.2 绘制 2D 图形 153
9.2 位图操作 156
9.3 Android 中的动画 158
9.3.1 Frame Animation 159
9.3.2 Tween Animation 161
9.3.3 Property Animation 168
9.3.4 AnimationListener（动画监听器） 168
9.4 在 Android 中播放音频与视频 169
9.4.1 MediaPlayer 介绍 169
9.4.2 运用 MediaPlayer 播放音频 171
9.4.3 播放视频 176
9.5 控制摄像头拍照 182

第 10 章 Android 网络编程基础 185
10.1 基于 TCP 协议的网络通信 185
10.1.1 TCP/IP 协议基础 185
10.1.2 使用 Socket 与 ServerSocket 建立通信 186
10.2 使用 URL 访问网络 190
10.2.1 使用 URL 获取网络资源 190
10.2.2 使用 URLConnection 提交请求 192
10.2.3 使用 HttpURLConnection 实现网络通信 197
10.3 使用 WebView 199
10.3.1 使用 WebView 浏览网页 199
10.3.2 加载本地 HTML 网页 201
10.3.3 JavaScript 交互调用 202
10.4 使用 WebService 进行网络编程 205
10.4.1 WebService 基础 205
10.4.2 调用 WebService 206
10.4.3 实现手机归属地查询 207

第 11 章 Android 数据存储 211
11.1 使用 SharedPreferences 211
11.1.1 SharedPreferences 的使用方法 211
11.1.2 SharedPreferences 的应用 212
11.2 File 存储 213
11.2.1 使用 I/O 流操作文件 213
11.2.2 文件操作应用 214
11.2.3 将文件保存到 SD 卡 216
11.3 SQLite 数据库 218
11.3.1 SQLite 数据库介绍 219
11.3.2 SQLite 数据库操作 221

第 12 章 GPS 应用开发 232
12.1 支持 GPS 的核心 API 232
12.2 获取 LocationProvider 233
12.3 获取定位信息 235

第 2 篇 Android 实验篇

实验 1 简单 UI 设计 240
1.1 实验目的 240
1.2 实验要求 240
1.3 实验内容 240
1.4 实验报告 245
1.5 实验成绩考核 246

实验 2 高级 UI 设计 247
2.1 实验目的 247
2.2 实验要求 247
2.3 实验内容 247
2.4 实验报告 255
2.5 实验成绩考核 256

实验 3 Intent 与 Activity 的使用 257
3.1 实验目的 257
3.2 实验要求 257
3.3 实验内容 257
3.4 实验报告 269
3.5 实验成绩考核 269

实验 4 Android 资源访问 270
4.1 实验目的 270

4.2	实验要求	270
4.3	实验内容	270
4.4	实验报告	277
4.5	实验成绩考核	277

实验 5 图形、图片与多媒体 278
　5.1　实验目的 278
　5.2　实验要求 278
　5.3　实验内容 278
　5.4　实验报告 284
　5.5　实验成绩考核 285

实验 6 Android 网络编程基础 286
　6.1　实验目的 286
　6.2　实验要求 286
　6.3　实验内容 286
　6.4　实验报告 295
　6.5　实验成绩考核 296

实验 7 SQLite 和 SQLiteDatabase 的使用 297
　7.1　实验目的 297
　7.2　实验要求 297
　7.3　实验内容 297
　7.4　实验报告 311
　7.5　实验成绩考核 312

附录 A "智能终端应用程序开发"在线金课 313
　A.1　课程访问方式 313
　A.2　在线金课课程体系 314
　A.3　教学内容 316
　A.4　教学方法 316

参考文献 318

第1篇
Android 基础篇

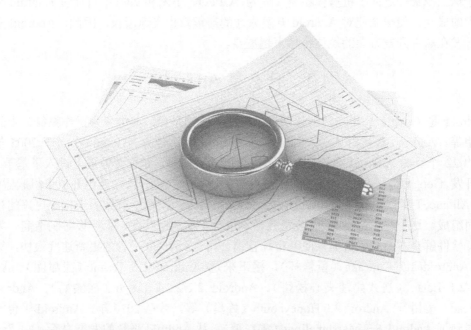

第 1 章　Android 系统与开发环境
第 2 章　Android UI 设计
第 3 章　基本程序单元 Activity
第 4 章　Android 应用核心 Intent 和 Filters
第 5 章　Android 事件处理
第 6 章　Android 服务
第 7 章　Android 广播接收器
第 8 章　ContentProvider 实现数据共享
第 9 章　图形、图片与多媒体
第 10 章　Android 网络编程基础
第 11 章　Android 数据存储
第 12 章　GPS 应用开发

第1章　Android 系统与开发环境

Android 是一种基于 Linux 内核（不包含 GNU 组件）的自由及开放源代码的操作系统，主要应用于移动设备，如智能手机和平板电脑。Android 系统也是一个可靠的平台，它可以经济、有效地安装、部署和提供支持，并且具有良好的设计、开发环境。如今，Android 系统已经成为全球应用最广泛的手机操作系统。华为、OPPO、小米和 vivo 手机通过 Android 平台获得了巨大的成功，使得企业对 Android 开发人才的需求量也飞速增长。因此，Android 平台上的移动开发在软件开发方面的地位将越来越重要。

1.1　什么是 Android

Android 是 Google 公司专门为移动设备提供支持的平台，其中包含操作系统、中间件和核心应用等。Android 最初由 Andy Rubin（Android 之父）创办。Google 公司于 2005 年收购了成立约 22 个月的 Android 公司，开启了短信、手机检索、定位等业务，进入了基于 Linux 平台的开发。Google 公司在 2007 年 11 月 5 日正式发布了这个平台，之后由开放手机联盟（Open Handset Alliance）组织进行开发。Open Handset Alliance 组织由一群致力于构建更好的移动电话的公司组成。这个组织由 Google 公司领导，包含了移动运营商、手持设备制造商、零部件制造商、软件解决方案和平台提供商及市场营销公司。Android 平台的更新速度很快，第一个版本是 Android 1.5 Cupcake（蛋糕杯），接下来是 Android 1.6 Donut（甜甜圈）、Android 2.0/2.0.1/2.1 Eclair（法式奶油夹心松饼）、Android 2.2/2.2.1 Froyo（冻酸奶）、Android 2.3 Gingerbread（姜饼）、Android 3.0 Honeycomb（蜂巢）等。截至 2016 年，Android 平台的版本已经发展到了 Android 6.0 Marshmallow（棉花糖）。从 Android 平台的发展来看，它已经不再局限于移动应用领域，还可以应用于可穿戴设备、大屏幕设备、汽车中控和物联网等平台。

1.1.1　Android 平台的架构

Android 平台是一种基于 Linux 的开放源代码的软件堆栈，为各类设备和机型而创建。图 1.1 为 Android 平台的主要组件。

1. Linux Kernel（Linux 内核）

Android 平台的基础是 Linux 内核。上层的 Android Runtime 就依靠 Linux 内核来执行底层功能，这些功能包括线程和低层内存管理等。Android 平台使用 Linux 内核可以让 Android 更加安全，并且允许设备制造商为 Linux 内核开发硬件驱动程序。

2. Hardware Abstraction Layer（HAL）

Hardware Abstraction Layer，即硬件抽象层，提供了 Android 平台的标准接口，用于向更高级别的 Java API 框架显示设备的硬件功能。HAL 包含多个库模块，其中每个模块都为特定类型的硬件组件实现了一个接口，如相机或蓝牙模块。当框架 API 要求访问设备的硬件时，Android 系统将为该硬件组件加载库模块。

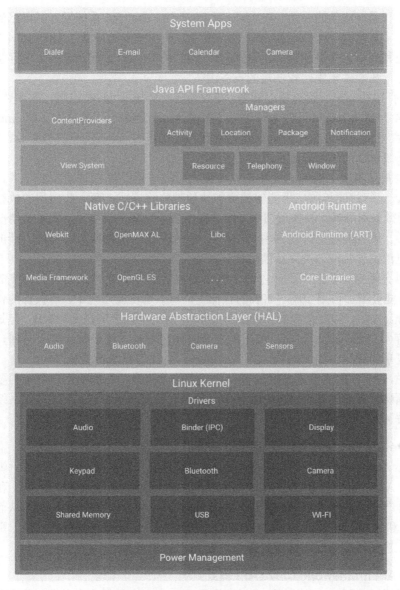

图 1.1　Android 平台的主要组件

3. Android Runtime

Android Runtime（ART）的机制与 Dalvik 不同。在 Dalvik 运行环境下，应用在每次运行时，都需要通过即时编译器将字节码转换为机器码，这会拖慢应用的运行效率。而在 ART 运行环境下，在第一次安装应用时，就会将字节码预先编译成机器码，使其成为真正的本地应用。这个过程称为预编译（Ahead-Of-Time，AOT）。这样一来，应用的启动（首次）和执行速度都会变得更快。

与 iOS 系统相比，Android 系统的用户体验有一个相对糟糕的开始。在很长的时间里，其界面一直像"丑小鸭"一样缺少美感，而且无法避免卡顿。不过，在 Google 公司的全力推动及硬件厂商的积极响应下，Android 系统还是跨越各种阻碍，逐渐壮大起来。

在此过程中，Google 公司也经历着重大的变化，逐渐从一个只重视数据的公司，转变为一个重视设计和用户体验的公司。从 Android 4.0 开始，Android 系统拥有了自己的设计语言和应用设计指导。与此同时，Google 公司也在着手解决系统卡顿问题。Android 4.1 的"黄油计划"使得系统和应用运行更加顺畅，而 Android 4.2 的 Project Svelte 加强了内存管理，使得系统能够顺利运行在硬件配置比较低端的设备上。

但是，所有这些都没有解决核心问题，即应用运行环境。Dalvik 的运行效率并不是最高的。从 Android 4.4 开始，Google 开发者就引进了新的 Android 运行环境 ART，即 Android Runtime（在 Android 官方页面的介绍中，也将其称为新的虚拟机），以替代旧的 Dalvik。然而 ART 是实验选项，系统默认的运行环境仍然是 Dalvik。

根据一些基准测试，新的运行环境能够使大多数应用的执行时间减少一半。这意味着，CPU 消耗大、运行时间长的应用能够被更加快速地执行完成，而一般应用的执行也能更加流畅，比如动画效果更顺畅，触控反馈更及时。在多核处理器的设备上，大多数情况下只需要激活少量的核心，或者能够更好地利用 ARM 的 big.LITTLE 架构即可。另外，新的运行环境将会显著提升电池的续航能力及系统的性能。

预编译也会带来一些缺点。一方面，机器码占用的存储空间更大。在将字节码转换为机器码之后，可能会增加占用 10%~20% 的存储空间，不过在应用包中，可执行的代码往往只是占用了一部分。比如，当前最新的 Google+ APK 占用的存储空间是 28.3MB，但是代码占用的存储空间只有 6.9MB。另一方面，应用的安装时间会变长。至于延长多少时间，取决于应用本身，一些复杂的应用（如 Facebook 和 Google+）会让用户等待更长时间。

总的来说，ART 的优点还是远远多于其缺点的。毕竟，在影响用户体验的要素中，电池续航和应用顺畅运行更为重要。在 2014 年 10 月 15 日发布的全新 Android 操作系统 Android 5.0 中，Dalvik 被 ART 彻底取代。

4. Native C/C++ Libraries（原生 C/C++库）

许多核心 Android 系统组件和服务（如 ART 和 HAL）都是基于以 C 和 C++编写的原生代码库构建的。Android 平台通过 Java 框架 API 可以向应用显示其中部分原生 C 和 C++库的功能。例如，用户可以通过 Android 框架的 Java OpenGL API 访问 OpenGL ES，从而在应用中绘制和操作 2D 和 3D 图形。因此，如果在开发时需要用到 C 或 C++代码的应用，那么可以使用 Android NDK 直接从原生代码访问某些原生 C 和 C++平台库。

5. Java API Framework（Java API 框架）

用户可以通过使用 Java 编写的 API 来实现 Android OS 的整个功能集。通过这些 API 创建 Android 应用所需的构建块，这种方法可以简化核心模块化系统组件和服务的重复使用。这些组件和服务包括如下内容。

- 丰富、可扩展的视图系统：可用于构建应用的 UI，包括列表、网格、文本框、按钮，甚至可嵌入式网络浏览器等。
- 资源管理器：用于访问一些非代码资源，如本地化的字符串、图形和布局文件等。
- 通知管理器：用于让所有应用在状态栏中显示自定义提醒。
- Activity 管理器：用于管理应用的生命周期，提供常见的导航返回栈等。
- 内容提供程序：用于让应用访问其他应用（如"联系人"应用）中的数据或者共享自己的数据。

6. System Apps（系统应用）

Android 系统随附了一套关于电子邮件、短信、日历、网络浏览器和联系人等的核心应用。

平台随附的应用与用户可以选择安装的应用一样，没有特殊性，因此第三方应用也可以成为用户的默认网络浏览器、短信 Messenger，甚至默认键盘等（有一些例外，如系统的"设置"应用）。这些系统应用可以被当作用户的应用，同时它们也可以为开发者自己的应用提供一些专属功能。比如，如果用户需要开发一个有发送短信功能的应用，那么他无须自己构建发送短信功能，直接通过调用短信应用向用户指定的接收者发送消息就可以了。

1.1.2 Android 系统的功能

1．显示布局

Android 系统支持更大的分辨率，VGA、2D 显示、3D 显示都将支持 OpenGL ES 3.0 标准规格（从 Android 4.3 开始支持 OpenGL ES 3.0），并且支持传统的智能手机。

2．数据存储

Android 系统内置 SQLite 小型关联式数据库管理系统，用来存储数据。

3．网络

Android 系统支持所有的网络制式，包括 GSM/EDGE、IDEN、CDMA、TD-SCDMA、5G、EV-DO、UMTS、Bluetooth、Wi-Fi、LTE、NFC 和 WiMAX。

4．信息

作为智能手机操作系统，Android 系统支持短信和邮件，并且支持所有的云信息和服务器信息。

5．语言

Android 系统支持多语言。

6．浏览器

Android 系统中内置的网页浏览器基于 WebKit 核心，并且采用了 Chrome V8 引擎。在 Android 4.0 内置的浏览器测试中，HTML5 和 Acid3 在故障处理中均获得了满分。Android 2.2～Android 4.0 支持 Flash，在 Android 4.0 之后不再支持 Flash。

7．支持 Java

虽然 Android 系统中的应用程序大部分都是使用 Java 编写的，但是 Android 系统是以转换为 Dalvik Executables 的文件在 Dalvik 虚拟机上运行的。由于 Android 系统中并不自带 Java 虚拟机，因此无法直接运行 Java 程序。不过，Android 平台上提供了多个 Java 虚拟机，以供用户下载使用，安装了 Java 虚拟机的 Android 系统可以运行 Java_ME 程序。Android 5.0（Lolipop）开始以 Android Runtime（ART）取代 Dalvik 虚拟机。

8．媒体支持

Android 系统本身支持以下格式的音频/视频/图片媒体：WebM、H.263、H.264（in 3GP or MP4 container）、MPEG-4 SP、AMR、AMR-WB（in 3GP container）、AAC、HE-AAC（in MP4 or 3GP container）、MP3、MIDI、Ogg Vorbis、FLAC、WAV、JPEG、PNG、GIF、BMP。若用户需要播放更多格式的多媒体文件，则可以安装其他第三方应用程序。

9．流媒体支持

Android 系统支持 RTP/RTSP（3GPP PSS、ISMA）的流媒体及 HTML5 <video>的流媒体，

同时支持 Adobe 的 Flash，在安装了 RealPlayer 之后，还支持苹果公司的流媒体。

10. 硬件支持

Android 系统支持视频/照片摄像头、多点电容/电阻触摸屏、GPS，支持加速计、陀螺仪、气压计、磁力仪（高斯计）、键盘、鼠标、USB Disk、专用的游戏控制器、体感控制器、游戏手柄、蓝牙设备、无线设备、感应和压力传感器、温度计，还支持加速 2D 位块传输（硬件方向、缩放、像素格式转换）和 3D 图形加速。

11. 多点触控

Android 系统支持本地的多点触控，在最初的 HTC Hero 智能手机上有这个功能。该功能是内核级别的（为了避免对苹果公司的触屏技术造成侵权）。

12. 蓝牙

Android 系统支持 A2DP、AVRCP、发送文件（OPP）、访问电话簿（PBAP）、语音拨号和发送智能手机之间的联系，还支持键盘、鼠标和操纵杆（HID）。

13. 多任务处理

Android 系统支持本地的多任务处理。

14. 语音功能

除了支持普通的电话通话，Android 系统从最初的版本开始就支持通过语音操作来使用 Google 进行网页搜索等功能。而从 Android 2.2 开始，语音功能还可以用来输入文字、语音导航等。

15. 无线共享功能

Android 系统支持用户使用本机充当"无线路由器"，并且将本机的网络共享给其他智能手机，其他机器只需要通过 Wi-Fi 找到共享的无线热点，就可以上网。而在 Android 2.2 之前的系统中，则需要通过第三方应用或其他定制版系统来实现这个功能。

16. 截图功能

从 Android 4.0 开始，系统可支持截图功能，该功能允许用户直接抓取智能手机屏幕上的任何画面，使得用户可以通过编辑功能对截图进行处理，并通过蓝牙、E-mail、微博、共享等方式发送给其他用户，或者上传到网络上，或者复制到计算机中。

1.1.3　Android 系统分支

- Phone and Tablet：Phone and Tablet 是运行在智能手机和平板电脑的一个 Android 系统分支。
- Wear OS：Wear OS 是穿戴式 Android 系统，可直接在手表上运行，并且可以访问传感器和 GPU 等硬件。开发穿戴式应用与开发使用 Android SDK 的其他应用相似，但在设计和功能上有所不同。
- Android TV：Android TV 是可以让用户在大屏幕（如电视）上体验沉浸式内容的 Android 系统分支。用户可以在主屏幕上发现内容推荐信息，由 Leanback 库提供各种 API，可以帮助用户打造绝佳的遥控器使用体验。
- Automotive：Automotive 系统分支可以帮助用户构建在行车过程中通过 Android Automotive OS 和 Android Auto 进行连接的应用。搭载 Android Automotive OS 的车辆的用户可以将

自己的应用安装到该车辆的信息娱乐系统中。而 Android Auto 可以让用户将手机（搭载 Android 5.0 或更高版本）连接到兼容的车辆上，以便直接在控制台上显示针对驾驶员进行了优化的应用版本。
- Android Things：Android Things 系统分支可以帮助用户构建智能的联网设备应用。

在创建项目时选择 Android 系统分支的应用界面如图 1.2 所示。

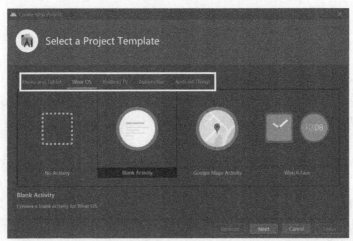

图 1.2　在创建项目时选择 Android 系统分支的应用界面

1.1.4　Android 平台五大优势特色

1．开放性

在优势方面，Android 平台最突出的就是其开放性，其开放的平台允许任何移动终端厂商加入 Android 联盟。显著的开放性可以使其拥有更多的开发者，随着用户和应用的日益丰富，即使是一个崭新的平台也将很快走向成熟。开放性对 Android 平台的发展而言，有利于该平台积累人气，这里的人气包括消费者和厂商；而对消费者而言，最大的受益是丰富的软件资源。开放的平台也会带来更大的竞争，如此一来，消费者将可以用更低的价格购得心仪的手机。

2．摆脱运营商的束缚

在过去很长的一段时间里，特别是在欧美地区，手机应用往往受到运营商的制约，如使用什么功能、接入什么网络，几乎都受到运营商的控制。自 iPhone 上市以来，用户可以更加方便地连接网络，逐渐摆脱运营商的束缚。随着 EDGE 和 HSDPA 这些 2G 到 3G 移动网络的逐步过渡和提升，手机随意接入网络已不是运营商口中的"笑谈"。

3．丰富的硬件选择

这一点还是与 Android 平台的开放性相关，由于 Android 的开放性，众多厂商会推出各具功能特色的产品。而功能上的差异和特色不会影响到数据同步甚至软件的兼容性，就好比用户从诺基亚的 Symbian 手机改用苹果的 iPhone 手机时，可以将 Symbian 手机中优秀的软件带到 iPhone 手机上使用，也可以将联系人等资料方便地转到 iPhone 手机上。

4．不受任何限制的开发商

由于 Android 平台提供给第三方开发商一个十分宽泛、自由的环境，因此开发商不会受到各种条条框框的限制。可想而知，这会使很多新颖、别致的软件诞生。但也有其两面性，如何控制不良程序和游戏的推出是留给 Android 的难题之一。

5. 无缝结合的 Google 应用

Google 公司已经拥有了 20 多年历史。从搜索巨人到全面的互联网渗透，Google 应用（如地图、邮件、搜索等）已经成为连接用户和互联网的重要纽带，而 Android 手机将无缝结合这些优秀的 Google 应用。

1.2 Android Studio 入门

Android Studio 是一套由 Google 公司推出的、以 IntelliJ IDEA 为基础的 Android 集成开发环境，目前已经免费向 Google 及 Android 的开发人员开放。为了简化 Android 平台的开发力度，Google 公司决定重点建设 Android Studio 工具。Google 公司已在 2015 年年底停止支持其他集成开发环境，如 Eclipse。

Android Studio 是第一个官方的 Android 开发环境。其他工具，如 Eclipse，在 Android Studio 发布之前已经有了大规模的使用。为了帮助开发者转向使用 Android Studio，Google 公司已经写出一套迁移指南。同时，Google 公司发布声明称，在接下来的时间里，会为 Android Studio 增加一些性能工具，Eclipse 里现有的 Android 工具也会通过 Eclipse 基金会继续支持下去。

1.2.1 Android Studio 的特点

- 可视化布局：包括 WYSIWYG 编辑器、实时编码、实时程序界面预览功能。
- 开发者控制台：包括优化提示、协助翻译、来源跟踪、宣传和营销曲线图、使用率度量功能。
- 支持基于 Gradle 的构建。
- 拥有 Android 特定代码重构和快速修复功能。
- Lint 提示工具可以更好地对程序性能、可用性、版本兼容和其他问题进行控制捕捉。
- 支持 ProGuard 和应用签名功能。
- 可基于模板的向导生成常用的 Android 应用设计和组件。
- 自带布局编辑器，可以让开发者拖放 UI 控件，并预览在不同尺寸设备上的 UI 显示效果等。
- 支持构建 Android Wear、TV 和 Auto 应用。
- 内置 Google Cloud Platform，支持 Google Cloud Messaging 和 App Engine 的集成。

1.2.2 Android Studio 系统要求

Android Studio 系统要求如表 1.1 所示。

表 1.1 Android Studio 系统要求

项 目	Windows	OS X	Linux
操作系统版本	Microsoft Windows 10/8.1/8/7/Vista/ 2003（32 或 64 位）	OS X 10.8.5 或更高版本，最高版本号为 10.10.5（Yosemite）	GNOME、KDE、Unity Desktop on Ubuntu、Fedora、GNU/Linux Debian
内存	最低 2GB，推荐 4GB 内存		
磁盘空间	500MB 磁盘空间		
Space for Android SDK	至少 1GB 用于 Android SDK、模拟器系统映像和缓存		

项　目	Windows	OS X	Linux
JDK 版本	Java Development Kit（JDK）7 或更高版本		
屏幕分辨率	最低 1280 像素×800 像素		

1.2.3　Android Studio 和 Eclipse ADT 的比较

Android Studio 和 Eclipse ADT 的比较如表 1.2 所示。

表 1.2　Android Studio 和 Eclipse ADT 的比较

特　性	Android Studio	Eclipse ADT
编译系统	Gradle	Ant
基于 Maven 的构建依赖	是	否
构建变体和多 APK 生成	是	否
高级的 Android 代码完成和重构	是	否
图形布局编辑器	是	是
APK 签名和密钥库管理	是	是
NDK 支持	Beta	是

1.3　Android 开发环境的搭建

1.3.1　JDK 的下载

JDK 是 Sun 公司（2009 年，Sun 被 Oracle 收购）针对 Java 开发人员发布的免费软件开发工具包（Software Development Kit，SDK）。自 Java 推出以来，JDK 已经成为使用最广泛的 Java SDK。作为使用 Java 编写的 SDK，普通用户并不需要安装 JDK 来运行 Java 程序，只需要安装 JRE（Java Runtime Environment）。而程序开发人员则必须安装 JDK 来编译、调试程序。下面以 JDK 15 为例，介绍下载 JDK 的方法，具体步骤如下。

（1）打开浏览器，输入网址，进入 Oracle 的官网主页，如图 1.3 所示。

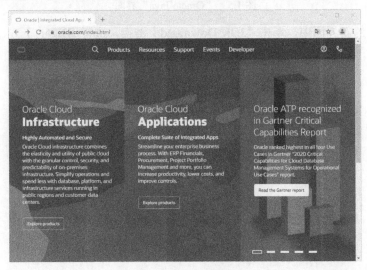

图 1.3　Oracle 的官网主页

（2）选择"Products"选项卡，选择"Java"选项，跳转到"Oracle Java"界面，单击"Download Java"按钮，如图 1.4 和图 1.5 所示。

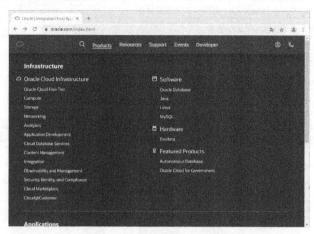

图 1.4　"Products"选项卡

图 1.5　"Oracle Java"界面

（3）在弹出的新界面中，单击"JDK Download"链接，进行 JDK 的下载，如图 1.6 所示。

图 1.6　JDK 下载界面

1.3.2 JDK 的安装

在下载完适合自己操作系统的 JDK 版本以后，就可以进行安装了。下面以 Windows 系统为例，讲解 JDK 的安装步骤。

（1）双击 JDK 安装包，会出现如图 1.7 所示的 JDK 安装向导，然后单击"下一步"按钮。

（2）在打开的如图 1.8 所示的界面中，单击"更改"按钮，将安装位置修改为"C:\Java\jdk-15.0.2\"（根据版本决定最后的 JDK 编号），并单击"确定"按钮，如图 1.9 所示。

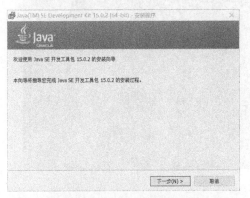
图 1.7 JDK 安装向导

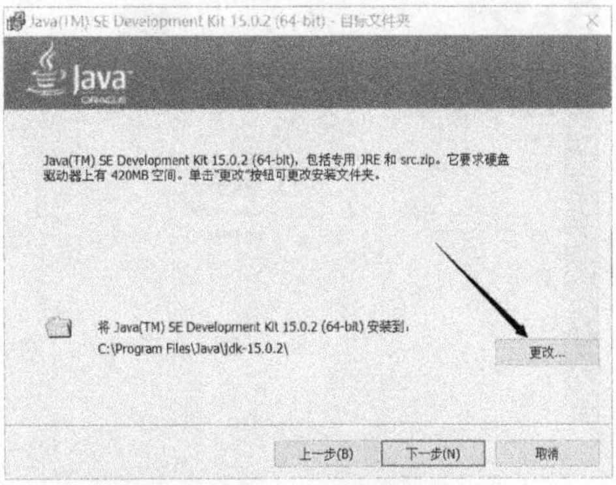

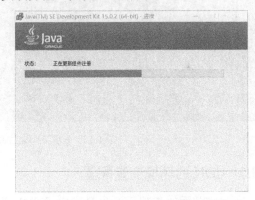
图 1.8 JDK 安装位置选择界面（1）

（3）单击"下一步"按钮，弹出 JDK 正在安装界面，如图 1.10 所示。

图 1.9 JDK 安装位置选择界面（2）

图 1.10 JDK 正在安装界面

（4）在安装完成后，弹出如图 1.11 所示的界面，单击"关闭"按钮，结束安装。

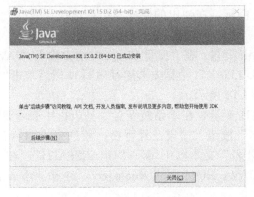
图 1.11 JDK 安装完成界面

（5）JDK 11 以上版本没有 JRE 的解决方法。
- 以管理员身份运行 CMD，打开"命令提示符"窗口并输入"cd C:\Java\jdk-15.0.2"（根据安装路径填写）。
- 使用"bin\jlink.exe --module-path jmods --add-modules java.desktop --output jre"命令手动生成 JRE，如图 1.12 所示。

图 1.12　JDK 11 以上版本没有 JRE 的解决方法

1.3.3　JDK 的环境变量配置

在完成前面的步骤后，只是完成了 JDK 环境的安装。这时还需要配置一系列的环境变量才能使用 JDK 环境进行 Android/Java 开发。配置的环境变量包括 JAVA_HOME、CLASSPATH 和 Path。

（1）右击"此电脑"图标，在弹出的快捷菜单中选择"属性"命令，并单击"高级系统设置"→"环境变量"→"新建"按钮，如图 1.13 所示。

（2）配置 JAVA_HOME 变量。在"变量名"文本框中输入"JAVA_HOME"，在"变量值"文本框中输入"C:\Java\jdk-15.0.2\"（根据安装路径填写），然后单击"确定"按钮，JAVA_HOME 变量就配置完成了，如图 1.14 所示。

（3）配置 CLASSPATH 变量。在"系统变量"列表框中查看是否有 CLASSPATH 变量，如果没有，就单击"新建"按钮；如果已经存在，就选择"CLASSPATH"选项，单击"编辑"按钮，然后在"变量名"文本框中填写"CLASSPATH"，在"变量值"文本框中添加";;%JAVA_HOME%\lib;%JAVA_HOME%\lib\tools.jar"。注意：添加部分与前面部分用";"隔开，如图 1.15 所示。

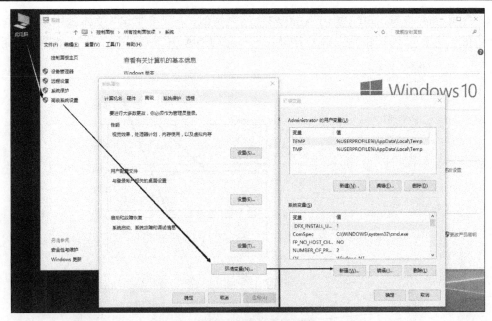

图 1.13 环境变量配置

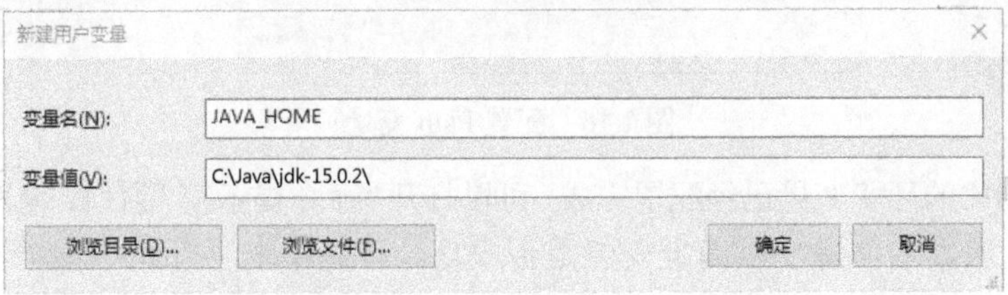

图 1.14 配置 JAVA_HOME 变量

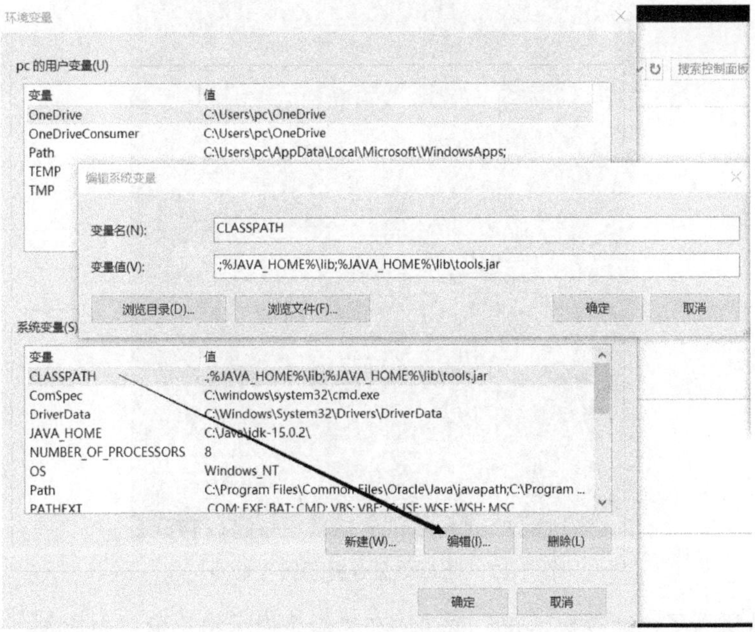

图 1.15 配置 CLASSPATH 变量

（4）配置 Path 变量。与配置 CLASSPATH 变量类似，在"变量名"文本框中填写"Path"，在"变量值"文本框中添加"%JAVA_HOME%\bin;%JAVA_HOME%\jre\bin"，如图 1.16 所示。

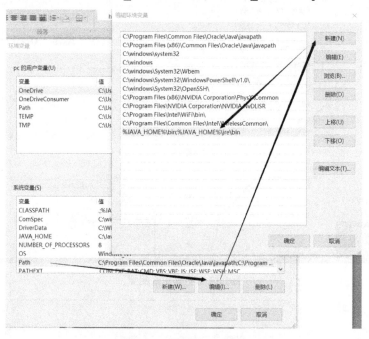

图 1.16　配置 Path 变量

（5）至此，JDK 的环境变量已经配置完成，可以打开"命令提示符"窗口，输入"Java -version"命令，查看 Java 版本的信息，以确定安装是否成功。首先单击"开始"菜单按钮，然后选择"所有应用程序"→"Windows 系统"→"命令提示符"命令，如图 1.17 所示。

图 1.17　选择"命令提示符"命令

（6）这时就进入了"命令提示符"窗口。在"命令提示符"窗口中输入"Java -version"命令。注意："Java"和"-version"之间有一个空格，然后按 Enter 键，如图 1.18 所示。

（7）JDK 版本信息就全部显示出来了，表明 JDK 已经安装和配置完成，可以开始进行 Java 程序开发了，如图 1.19 所示。

图 1.18　在"命令提示符"窗口中运行"Java -version"命令

图 1.19　JDK 版本信息

1.3.4　在 Windows 环境下安装 Android Studio

1．准备工具

（1）JDK 安装包（JDK 9 及更高版本）。

（2）Android Studio 安装文件。

2．下载安装文件

（1）android-studio-ide-201.7042882-windows.exe（推荐）。

（2）android-studio-ide-201.7042882-windows.zip（解压版本）。

Android Studio 下载选项如图 1.20 所示（Android 官方网站会实时更新最新安装文件，请根据最新安装文件确定安装方式）。

Android Studio downloads

Platform	Android Studio package	Size	SHA-256 checksum
Windows (64-bit)	android-studio-ide-201.7042882-windows.exe Recommended	896 MiB	22cdcfeffabe384e788b679c5c03238684fa3f1f4c73e2a3744f2fe5aab7f97f
	android-studio-ide-201.7042882-windows.zip No .exe installer	900 MiB	7fb6e49e76ead2ff389c37f83a6c90526aa1f716aece028c8b8c34edf8ce9804
Mac (64-bit)	android-studio-ide-201.7042882-mac.dmg	877 MiB	541db2ab0fda0b1197509b39fac905b7e4879a1d0bad749ad1ccc0727e02ea6b
Linux (64-bit)	android-studio-ide-201.7042882-linux.tar.gz	882 MiB	89f7c3a03ed928edeb7bbb1971284bcb72891a77b4f363557a7ad4ed37652bb9
Chrome OS	android-studio-ide-201.7042882-cros.deb	742 MiB	13a7bda7a58cd56e1544f16705a17cc633951d692a16c0b9a9767b07d7cfea54

See the Android Studio release notes. More downloads are available in the download archives.

图 1.20　Android Studio 下载选项

3．安装

（1）找到下载的安装文件。

（2）双击该文件，开始安装。欢迎安装界面如图 1.21 所示。

（3）单击"Next"按钮，进入组件选项界面，选择需要安装的 Android Studio 功能，如图 1.22 所示。

图 1.21　欢迎安装界面

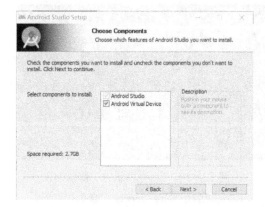
图 1.22　组件选项界面

（4）单击"Next"按钮，选择 Android Studio 的安装路径，如图 1.23 所示。

（5）单击"Next"按钮，设置快捷方式，如图 1.24 所示。

图 1.23　选择安装路径

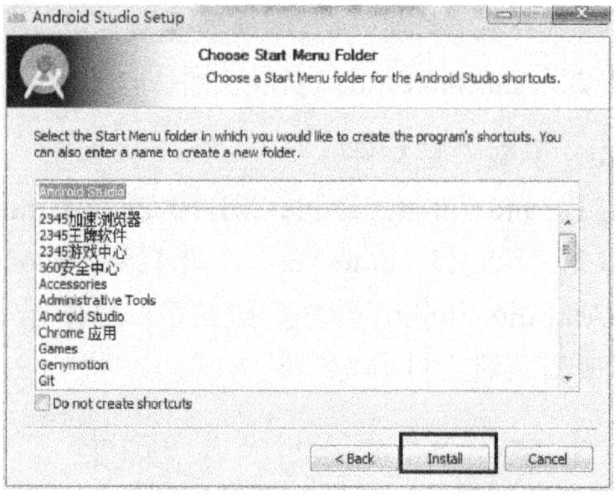

图 1.24　设置快捷方式

（6）单击"Install"按钮，开始安装，并显示安装进度，如图 1.25 所示。

图 1.25　显示安装进度

（7）在安装完成后，单击"Finish"按钮，完成安装，如图 1.26 所示。
（8）Android Studio 启动界面如图 1.27 所示。

图 1.26　安装完成

图 1.27　Android Studio 启动界面

（9）Android Studio 在第一次启动后，会弹出指定 SDK 路径的选项界面，如图 1.28 所示。

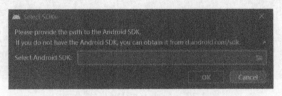

图 1.28　指定 SDK 路径的选项界面

（10）如果本机上从来没有安装、下载过 SDK，则单击图 1.28 中的"Cancel"按钮和弹出界面中的"否"按钮，取消指定 SDK 路径；如果本机上有提前下载好的 SDK，则把路径指向 SDK 下载、保存后的路径即可。

1.3.5　下载、安装和配置 Android SDK

SDK Tools 其实就是 Android SDK Manager，是用于管理各种版本的 SDK 的工具。在 Android SDK 中，包含模拟器、教程、API 文档和示例代码等内容。下面以 Windows 为例详细讲解下载和安装 Android SDK 的步骤。

（1）选择"File"→"Settings"命令，如图 1.29 所示，打开 Android Studio 设置界面。

图 1.29　选择"Settings"命令

（2）在设置界面的搜索栏中输入"sdk"，快速找到 Android SDK 设置界面，如图 1.30 所示。

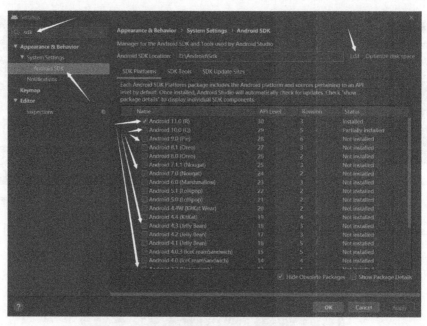

图 1.30　Android SDK 设置界面

（3）在 Android SDK 设置界面中勾选需要下载的 SDK 版本和 SDK Tools 版本对应的复选框，然后单击"Apply"按钮。若系统报错，如出现"Your Android SDK is missing, out of date or corrupted"问题，则单击"Edit"按钮，将弹出如图 1.31 所示的 SDK 下载界面，再单击"Next"按钮，即可进行下载、安装，如图 1.32 所示。

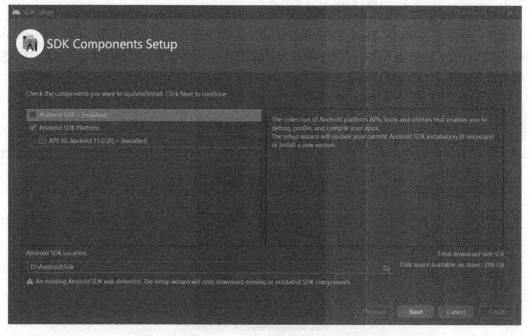

图 1.31　SDK 下载界面（1）

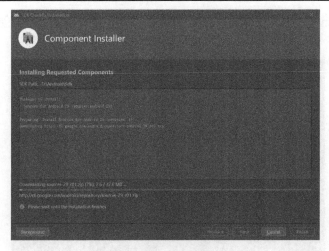

图 1.32　SDK 下载界面（2）

（4）在 Android Studio 启动界面右下角会出现如图 1.33 所示的界面，请耐心等待下载/加载完毕再进行编程活动。

图 1.33　Android Studio 下载/加载进度条界面

1.4　Android 自带模拟器 AVD

1.4.1　什么是 AVD

Android Studio 为开发者提供了可以在计算机上运行的虚拟手机，Android 将其称为 Android Virtual Device（AVD）。如果用户没有 Android 手机，则可以在 AVD 上运行各种 App 和自己开发的应用。

1.4.2　通过 AVD Manager 创建 AVD

（1）单击 Android Studio IDE 界面中的"AVD Manager"按钮，如图 1.34 所示，打开 AVD Manager 界面。

图 1.34　单击"AVD Manager"按钮

（2）在 AVD Manager 界面中单击中间的"Create Virtual Device"按钮，创建自定义的 AVD，如图 1.35 所示。

图 1.35　AVD Manager 界面

（3）在弹出的选择硬件设备界面中选择一款合适的 AVD，然后单击"Next"按钮，如图 1.36 所示。

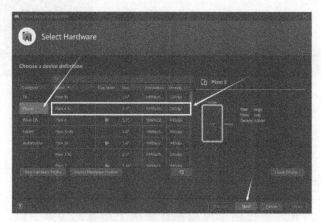

图 1.36　选择硬件设备界面

（4）在 AVD 下载界面中，根据 API 版本选择需要的 AVD 版本，单击"Download"链接进行下载，如图 1.37 所示。

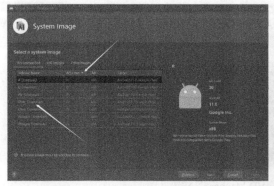

图 1.37　AVD 下载界面

（5）在 AVD 下载完毕后，可以自定义 AVD 的名称，然后单击"Finish"按钮，如图 1.38 所示。

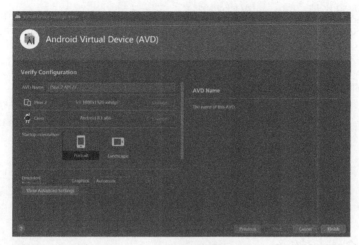

图 1.38　自定义 AVD 的名称

（6）在 AVD 列表中选择已下载的 AVD，单击  按钮，启动 AVD，如图 1.39 所示。
（7）在 AVD 运行成功后，将出现 Android 模拟器运行界面，如图 1.40 所示。

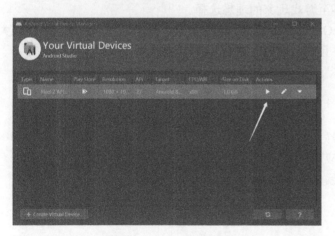

图 1.39　AVD 启动界面　　　　　　　　图 1.40　Android 模拟器运行界面

注意：如果 Android SDK 和 Android Studio 没有安装在 C 盘默认路径下，但 AVD Manager 在下载和启动 AVD 时会将其保存在默认路径中（C:\Users\Administrator\.android\avd），并不是修改后的路径中，这就会导致 AVD 启动报错（The emulator process for AVD (xxx) was killed）。修改方法如下。

模拟器在启动时，会默认按照以下顺序查找 AVD 目录：

$ANDROID_AVD_HOME
$ANDROID_SDK_HOME/.android/avd/
$HOME/.android/avd/

所以开发者需要检查环境变量的配置是否正确，分别添加 ANDROID_SDK_HOME 和 ANDROID_AVD_HOME 系统变量，如图 1.41 所示。

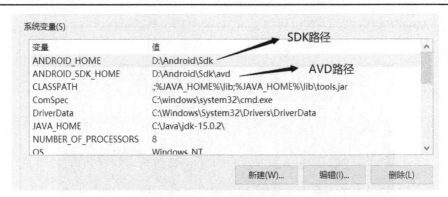

图 1.41　添加系统变量

1.5　Android 模拟器 Genymotion

1.5.1　什么是 Genymotion

由于国内访问 Google 并不便利，因此无论是离线下载还是在线下载，安装 Android 模拟器都变得非常困难。Genymotion 工具提供了一套完整的 Android 虚拟环境，是使用户迅速成为开发人员、测试人员、推销人员甚至游戏玩家的理想选择。

Genymotion 支持 Windows、macOS、Linux 系统，易于安装、易于运行。它通过功能强大的传感器来测试用户的应用程序，并且完美地融入整个开发环境中。

1.5.2　Genymotion 的特性

1．最好的 Android 模拟体验

（1）支持 OpenGL 加速，提供最好的 3D 性能体验。
（2）可以通过 Google Play 安装应用。
（3）支持全屏并改善了使用体验。

2．全控制

（1）可同时启动多个模拟器。
（2）支持传感器管理，如电池状态、GPS、Accelerator 加速器。
（3）支持 Shell 控制模拟器。
（4）完全兼容 ADB，用户可以通过主机控制其模拟器。

3．管理设备

（1）易安装。
（2）兼容 Microsoft Windows 32/64bit、Mac OS X 10.5 和 Linux 32/64bit。
（3）可以配置模拟器参数，如屏幕分辨率、内存大小、CPU 数量。
（4）轻松下载、部署最新的 Genymotion 虚拟设备。

4．通过 Android Studio 启动虚拟设备

使用 Genymotion 测试用户的应用。

1.5.3 Genymotion 运行环境要求

1. 需要满足以下操作系统之一
（1）Microsoft Windows Vista（32/64bit）及更高版本。
（2）Linux Ubuntu 12.04（32/64bit）及更高版本。
（3）Linux Debian Wheezy 64bit 及更高版本。
（4）Mac OS X 10.8（64bit）及更高版本。

2. 系统环境要求
（1）支持 OpenGL 2.0。
（2）CPU 支持 VT-x 或 AMD-V 虚拟化，通过 BIOS 设置开启。
（3）至少 2GB 内存。
（4）至少提供 400MB 的硬盘空间来安装 Genymotion，至少提供 2GB 的硬盘空间来部署 Genymotion 虚拟设备。有些可能需要 8GB 以上，这主要取决于用户安装在虚拟机中的应用程序。
（5）可用网络连接（用于安装和更新）。
（6）系统显示分辨率不低于 1024 像素×768 像素。

3. 应用程序要求
Oracle VirtualBox 4.1 或更高版本（版本越高越好）。

1.5.4 注册和下载 Genymotion 模拟器

（1）注册 Genymotion 账号：打开浏览器，在地址栏中输入官网地址，进入如图 1.42 所示的 Genymotion 官网主页，单击界面右上角的 "Sign in" 按钮，进入登录界面。

图 1.42　Genymotion 官网主页

（2）单击 "Account an creation" 按钮，在弹出的 Genymotion 注册界面中创建个人账号，依次填写用户名、电子邮箱、密码、公司规模（可选项）、用途类型（可选项），并勾选最后一项 "I accept terms of the privacy statement"（同意条款）复选框，如图 1.43 所示。

（3）在注册成功后，将弹出如图 1.44 所示的 Genymotion 注册成功界面，显示了用户注册的 Genymotion 账号（用户注册时使用的 E-mail 地址）。同时，在注册成功后，系统会发送一封邮件到用户的电子邮箱，用户必须打开邮件，以完成验证。

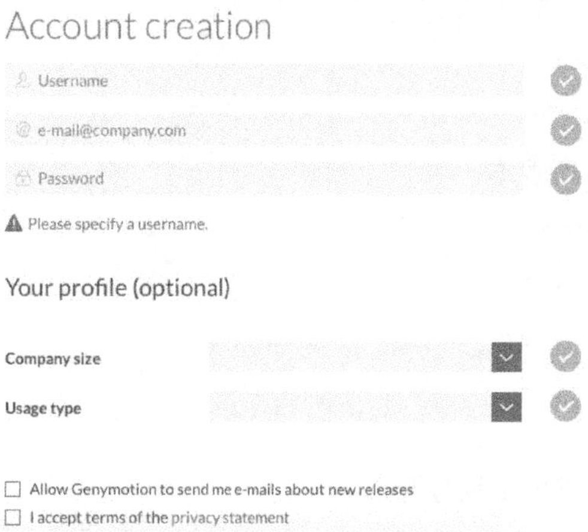

图 1.43　Genymotion 注册界面 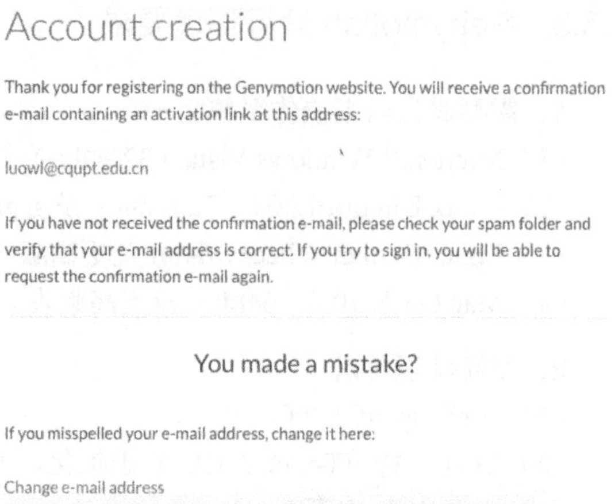　　　　图 1.44　Genymotion 注册成功界面

（4）回到 Genymotion 官网主页，如图 1.45 所示。单击界面中间的"Choose a plan"按钮，进入下载用户类型选择界面，如图 1.46 所示。

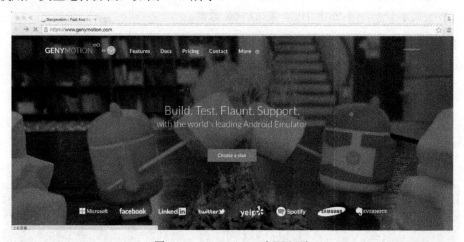

图 1.45　Genymotion 官网主页

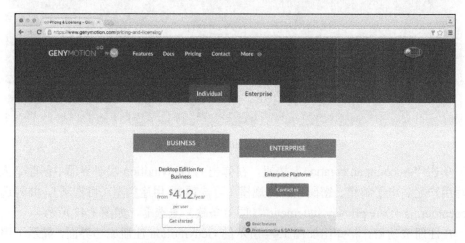

图 1.46　下载用户类型选择界面

（5）在当前界面中，选择"Individual"选项卡，进入个人下载界面，如图 1.47 所示。

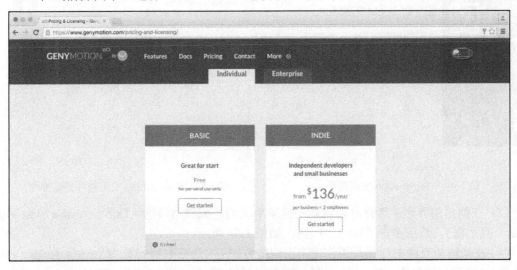

图 1.47　个人下载界面

（6）Genymotion 对于个人学习、使用是免费的，对于专业个人和两人以上的小型公司将收取 136 美元/年的费用。单击"BASIC"中的"Get started"按钮，进行免费下载。

（7）在如图 1.48 所示的下载界面中，用户根据个人计算机硬件和操作系统类型选择合适的安装文件。下面以 Windows 系统为例进行讲解。对于计算机中没有安装过 VirtualBox 的用户，建议下载 with VirtualBox 版本，在图 1.48 中单击"Download for Windows"按钮完成下载。

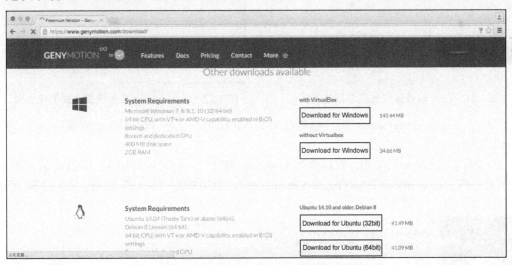

图 1.48　选择合适的安装文件

1.5.5　安装 Genymotion 模拟器

（1）双击下载的 Genymotion 安装文件，选择使用的语言并单击"Next"按钮，如图 1.49 所示。

（2）单击"Browse"按钮，可以更改安装路径，也可以使用软件默认的安装路径，即"C:\Program Files\Genymobile\Genymotion"，然后单击"Next"按钮，如图 1.50 所示。

 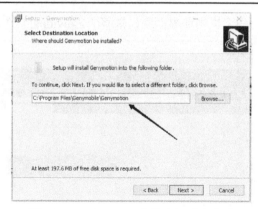

图 1.49　Genymotion 安装向导　　　　图 1.50　Genymotion 安装路径更改界面

（3）在弹出的界面中选择是否创建快捷菜单（勾选或取消勾选"Don't create a Start Menu folder"复选框），然后单击"Next"按钮，如图 1.51 所示。

（4）在弹出的界面中选择是否创建桌面快捷方式（勾选或取消勾选"Create a desktop icon"复选框），然后依次单击"Next"→"Install"→"Finish"按钮，如图 1.52 所示。

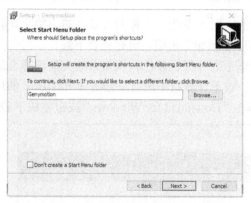 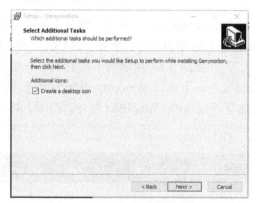

图 1.51　Genymotion 快捷菜单创建界面　　　　图 1.52　Genymotion 桌面快捷方式创建界面

（5）在安装完 Genymotion 后，会继续安装 VirtualBox。在 VirtualBox 安装向导中，单击"Next"按钮，如图 1.53 所示。

（6）单击"Browse"按钮，可以更改 Location 的地址，也可以使用 VirtualBox 软件默认的安装路径，即"C:\Program Files\Oracle\VirtualBox\"，然后单击"Next"按钮，如图 1.54 所示。

 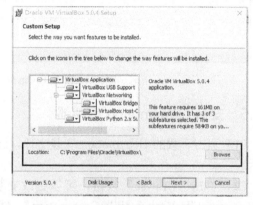

图 1.53　VirtualBox 安装向导　　　　图 1.54　VirtualBox 安装路径更改界面

（7）弹出 VirtualBox 确认安装界面，询问是否现在安装，单击"Yes"按钮，如图 1.55 所示。
（8）单击"Install"按钮，开始安装，如图 1.56 所示，然后单击"Finish"按钮，完成安装。

图 1.55　VirtualBox 确认安装界面　　　　图 1.56　开始安装 Genymotion

1.5.6　使用 Genymotion 模拟器

（1）在第一次进入 Genymotion 时，系统会检查用户是否安装了 Android 虚拟设备。如果没有安装，则会弹出对话框，询问用户是否现在添加一个虚拟设备，如图 1.57 所示，单击"Yes"按钮即可。

（2）在添加一个新的虚拟设备时，需要在 Genymotion 登录界面中输入用户名和密码进行验证，如图 1.58 所示。注意：若验证不通过，则可以到邮箱确认是否已经验证过。

图 1.57　第一次启动 Genymotion　　　　图 1.58　Genymotion 登录界面

（3）在验证成功后，可以看到有很多虚拟设备，如 Samsung Galaxy S3、S4 等，如图 1.59 所示。选择想添加的虚拟设备，然后单击"Next"按钮。

（4）开始下载虚拟设备，如图 1.60 所示，当下载到 100%时单击"Finish"按钮。

（5）回到主界面，选择一个已经添加的虚拟设备（如 Genymotion 模拟器），单击"Start"按钮，启动虚拟设备，如图 1.61 所示。

（6）启动 Genymotion 模拟器，将弹出如图 1.62 所示的手机模拟界面。

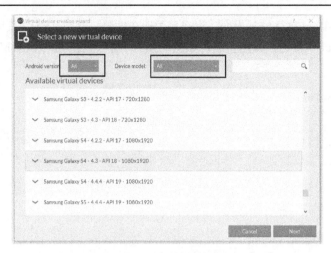

图 1.59　虚拟设备选择界面

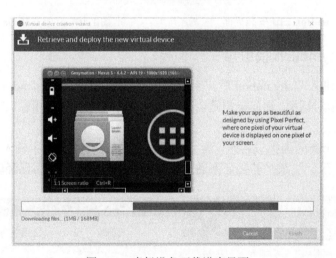

图 1.60　虚拟设备下载进度界面

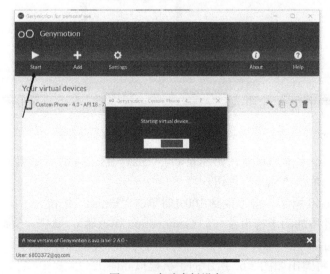

图 1.61　启动虚拟设备

图 1.62　手机模拟界面

1.5.7 给 Android Studio 安装 Genymotion 插件

（1）打开 Android Studio，选择"File"→"Settings"命令，如图 1.63 所示。

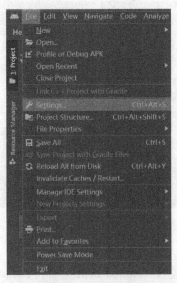

图 1.63　Android Studio 的"File"菜单

（2）在打开的界面中找到"Plugins"设置项，打开"Marketplace"选项卡，并在搜索框中输入"genymotion"关键词，单击"Installed"按钮，即可开始安装 Genymotion 插件，如图 1.64 所示。

图 1.64　Android Studio Plugins 设置界面

1.6　新建一个 HelloWorld 项目

在配置好 JDK 和 Android SDK 后，就可以开始新建 Android 项目了，具体步骤如下。

（1）启动 Android Studio，打开欢迎界面，选择"Create New Project"选项，如图 1.65 所示。

（2）在弹出的界面中选择"Empty Activity"选项，然后单击"Next"按钮，如图 1.66 所示。

图 1.65　在欢迎界面中新建一个 Android Studio 项目

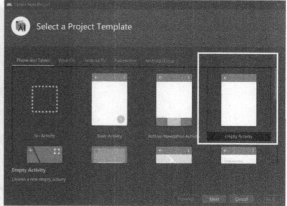

图 1.66　项目模板选择界面

（3）在"Name"文本框中输入"HelloWorld"，在"Package name"文本框中输入"cqupt.wenlong.helloworld"（根据自己需要改为姓名拼音.com 或公司英文名称.com），在"Save location"中设置项目保存路径，在"Language"下拉列表中选择"Java"选项，在"Minimum SDK"下拉列表中选择运行 Android 的最低 SDK 版本［此处以 API 19:Android 4.4（KitKat）为例］。最下面的"Use legacy android.support libraries"选项含义是使用旧库创建项目，此时不建议勾选。然后单击"Finish"按钮，如图 1.67 所示。

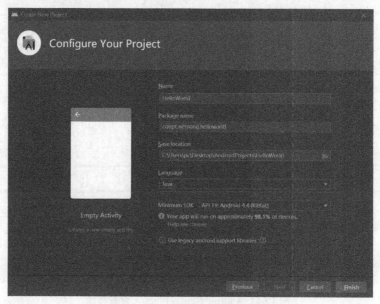

图 1.67　设置新项目名称界面

（4）在创建好项目后，会出现如图 1.68 所示的 Android Studio 编辑界面。

（5）在运行项目前，提前启动 AVD 模拟器或 Genymotion 模拟器。

（6）在模拟器平台启动成功后，会弹出如图 1.69 或图 1.70 所示的手机模拟界面。

第 1 章　Android 系统与开发环境

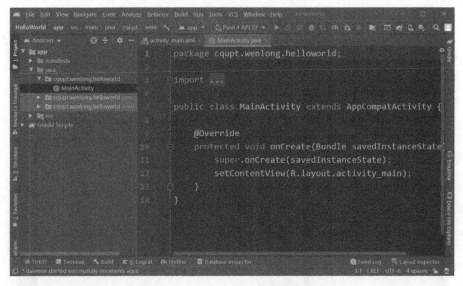

图 1.68　Android Studio 编辑界面

图 1.69　手机模拟界面（1）

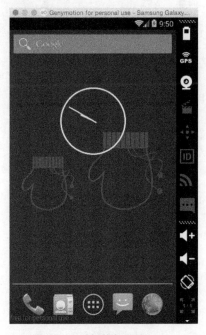

图 1.70　手机模拟界面（2）

（7）在 Android Studio 工具栏 ▶ 按钮左侧的下拉列表中可以选择计算机中正在运行的模拟器，并以此作为程序运行安装的平台，如图 1.71 所示。

图 1.71　正在运行的模拟器选项

（8）在选择好模拟器后，单击 Android Studio 工具栏中的 ▶ 按钮，运行程序。

（9）在模拟器中查看 HelloWorld 项目的运行效果，图 1.72 和图 1.73 分别展示了 HelloWorld 项目在 AVD 模拟器和 Genymotion 模拟器上的运行效果。

图 1.72　HelloWorld 项目的运行效果（1）

图 1.73　HelloWorld 项目的运行效果（2）

第 2 章　Android UI 设计

2.1　Android UI 布局

Android UI 布局一般使用布局管理器、ListView（列表视图）和 GridView（网格视图）3 种控件，下面对它们分别进行详细介绍。

Android 布局管理器可以很方便地控制各个控件的位置和大小，同时根据屏幕大小，管理容器内的控件，自动适配控件在手机屏幕中的位置。Android Studio 中提供了如图 2.1 所示的 7 种布局管理器：ConstraintLayout（约束布局管理器）、LinearLayout（horizontal）（水平线性布局管理器）、LinearLayout（vertical）（垂直线性布局管理器）、FrameLayout（帧布局管理器）、TableLayout（表格布局管理器）、TableRow（表格行布局管理器）、Space（空隙布局管理器）。这 7 种布局管理器的布局如图 2.2 所示。

图 2.1　Android Studio 中的 7 种布局管理器

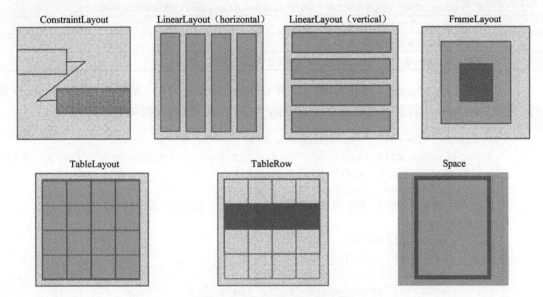

图 2.2　7 种布局管理器的布局

在一个 Android 应用程序中，用户界面是通过 View 和 ViewGroup 对象构建的。Android 中有很多种 View 和 ViewGroup 对象，它们都继承自 View 类。View 对象是 Android 平台上表示用户界面的基本单元。布局方式用于决定一组 View 对象如何布局，准确地说，它用于决定一个 ViewGroup 对象中包含的一些 View 对象是如何布局的。这里介绍的用于决定 View 对象的布局方式的类都直接或间接地继承自 ViewGroup 类，如图 2.3 所示。

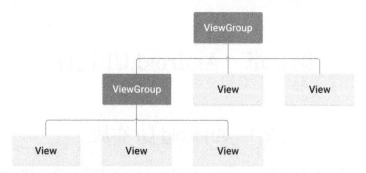

图 2.3　ViewGroup 对象如何在布局中形成分支并包含 View 对象的图示

其实，所有的布局方式都可以归类为 ViewGroup 类的 5 个类别，即 ViewGroup 类的 5 个直接子类。其他的一些布局都扩展自这 5 个类。

【例 2.1】　FrameLayout（帧布局管理器）实例

这个例子将使用 FrameLayout 创建一个简单的 Android 应用，下面将通过表 2.1 中的步骤来实现。

表 2.1　FrameLayout 实例步骤

步　骤	描　　　述
1	打开 Android Studio，创建一个 Android 应用，选择"Empty Activity"选项，在"Name"文本框中输入项目名，在"Package name"文本框中输入包名
2	在"Minimum SDK"下拉列表中选择"API 18: Android 4.3（Jelly Bean）"选项或其他 SDK
3	在"Language"下拉列表中选择"Java"选项
4	选择保存路径，单击"Finish"按钮
5	启动 Genymotion 模拟器，然后在 Android 工程项目中进行如下代码修改

将工程项目 res/mipmap 目录下的图片文件 ic_launcher.png 复制、粘贴到 res/drawable 目录下，并修改 res/layout 目录下的布局文件 activity_main.xml，修改代码如下：

```
<?xml version="1.0" encoding="utf-8"?>
<FrameLayout xmlns:android="http://schemas.android.com/apk/res/android"
    android:layout_width="fill_parent"
    android:layout_height="fill_parent">
<!--添加居中显示的Android图标图片ImageView，将显示在最下层-->
    <ImageView
        android:src="@drawable/th"
        android:scaleType="fitCenter"
        android:layout_height="392dp"
        android:layout_width="match_parent"
        android:layout_gravity="center_vertical" />
<!--添加居中显示的"Frame Demo"文字TextView，将显示在最上层-->
    <TextView
        android:text="Frame Demo"
        android:textSize="30px"
        android:textStyle="bold"
        android:layout_height="fill_parent"
        android:layout_width="fill_parent"
```

```
            android:gravity="center"
            android:layout_gravity="left|bottom" />
</FrameLayout>
```

程序在 Genymotion 中的运行效果如图 2.4 所示。

图 2.4　FrameLayout 实例运行效果

【例 2.2】　LinearLayout（horizontal/vertical）（水平/垂直线性布局管理器）实例

这个例子将使用 LinearLayout（horizontal/vertical）创建一个简单的 Android 应用，下面将通过表 2.2 中的步骤来实现。

表 2.2　LinearLayout（horizontal/vertical）实例步骤

步　　骤	描　　述
1	打开 Android Studio，创建一个 Android 应用，选择"Empty Activity"选项，在"Name"文本框中输入项目名，在"Package name"文本框中输入包名
2	在"Minimum SDK"下拉列表中选择"API 18: Android 4.3（Jelly Bean）"选项或其他 SDK
3	在"Language"下拉列表中选择"Java"选项
4	选择保存路径，单击"Finish"按钮
5	启动 Genymotion 模拟器，然后在 Android 工程项目中进行如下代码修改

修改工程项目的 res/layout 目录下的布局文件 activity_main.xml，修改代码如下：

```
<?xml version="1.0" encoding="utf-8"?>
<LinearLayout xmlns:android="http://schemas.android.com/apk/res/android"
    android:layout_width="fill_parent"
    android:layout_height="fill_parent"
    android:orientation="vertical" >
    <!--添加 START_SERVICE 按钮-->
    <Button android:id="@+id/btnStartService"
        android:layout_width="142dp"
        android:layout_height="wrap_content"
```

```
            android:text="START_SERVICE" />
        <!--添加 PAUSE_SERVICE 按钮-->
        <Button android:id="@+id/btnPauseService"
            android:layout_width="142dp"
            android:layout_height="wrap_content"
            android:text="PAUSE_SERVICE" />
        <!--添加 STOP_SERVICE 按钮-->
        <Button android:id="@+id/btnStopService"
            android:layout_width="142dp"
            android:layout_height="wrap_content"
            android:text="STOP_SERVICE" />

</LinearLayout>
```

程序在 Genymotion 中的运行效果如图 2.5 所示。

将 res/layout 目录下的布局文件 activity_main.xml 中的"android:orientation="vertical""修改为"android:orientation="horizontal"",再运行一次这个项目,单击模拟器的◇按钮,即可使手机屏幕转为横向,运行效果如图 2.6 所示。

图 2.5　LinearLayout（vertical）
　　　　实例运行效果

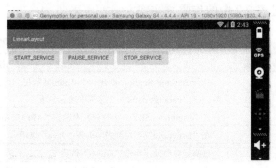

图 2.6　LinearLayout（horizontal）
　　　　实例运行效果

【例 2.3】　TableLayout（表格布局管理器）、TableRow（表格行布局管理器）实例

这个例子将使用 TableLayout、TableRow 创建一个简单的 Android 应用,下面将通过表 2.3 中的步骤来实现。

表 2.3　TableLayout、TableRow 实例步骤

步　骤	描　述
1	打开 Android Studio,创建一个 Android 应用,选择"Empty Activity"选项,在"Name"文本框中输入项目名,在"Package name"文本框中输入包名

步 骤	描 述
2	在"Minimum SDK"下拉列表中选择"API 18: Android 4.3（Jelly Bean）"选项或其他 SDK
3	在"Language"下拉列表中选择"Java"选项
4	选择保存路径，单击"Finish"按钮
5	启动 Genymotion 模拟器，然后在 Android 工程项目中进行如下代码修改

修改工程项目的 res/layout 目录下的布局文件 activity_main.xml，修改代码如下：

```xml
<?xml version="1.0" encoding="utf-8"?>
<TableLayout xmlns:android="http://schemas.android.com/apk/res/android"
    android:layout_width="fill_parent"
    android:layout_height="fill_parent">

    <TableRow
        android:layout_width="fill_parent"
        android:layout_height="fill_parent">
        <TextView
            android:text="Time"
            android:layout_width="wrap_content"
            android:layout_height="wrap_content"
            android:layout_column="1" />
        <TextClock
            android:layout_width="wrap_content"
            android:layout_height="wrap_content"
            android:id="@+id/textClock"
            android:layout_column="2" />
    </TableRow>

    <TableRow>
        <TextView
            android:text="First Name"
            android:layout_width="wrap_content"
            android:layout_height="wrap_content"
            android:layout_column="1" />
        <EditText
            android:width="200px"
            android:layout_width="wrap_content"
            android:layout_height="wrap_content" />
    </TableRow>

    <TableRow>
        <TextView
            android:text="Last Name"
            android:layout_width="wrap_content"
            android:layout_height="wrap_content"
            android:layout_column="1" />
        <EditText
            android:width="100px"
            android:layout_width="wrap_content"
```

```
            android:layout_height="wrap_content" />
    </TableRow>

    <TableRow
        android:layout_width="fill_parent"
        android:layout_height="fill_parent">
        <RatingBar
            android:layout_width="wrap_content"
            android:layout_height="wrap_content"
            android:id="@+id/ratingBar"
            android:layout_column="2" />
    </TableRow>

    <TableRow
        android:layout_width="fill_parent"
        android:layout_height="fill_parent" />
    <TableRow
        android:layout_width="fill_parent"
        android:layout_height="fill_parent">
        <Button
            android:layout_width="wrap_content"
            android:layout_height="wrap_content"
            android:text="SUBMIT"
            android:id="@+id/button"
            android:layout_column="2" />
    </TableRow>
</TableLayout>
```

程序在 Genymotion 中的运行效果如图 2.7 所示。

图 2.7　TableLayout、TableRow 实例运行效果

ConstraintLayout（约束布局管理器）是 Android Studio 2.2 中主要的新增功能之一。在传统的 Android 应用程序开发中，界面基本都是依靠编写 XML 代码完成的。虽然 Android Studio 也

支持通过可视化的方式来编写界面,但是实际操作起来并不方便,因此并不推荐使用可视化的方式来编写 Android 应用程序的界面。

而 ConstraintLayout 就是为了解决这个问题而出现的。它和编写界面的传统方式恰恰相反,ConstraintLayout 非常适合使用可视化的方式来编写界面,但并不太适合使用 XML 的方式来编写界面。当然,可视化操作的背后仍然还是使用 XML 代码来实现的,只不过这些代码是由 Android Studio 根据用户的操作自动生成的。

另外,ConstraintLayout 还有一个优点,它可以有效地解决布局嵌套过多的问题。用户在编写界面时,复杂的布局总会伴随着多层的嵌套,而嵌套越多,程序的性能就越差。ConstraintLayout 是使用约束的方式来指定各个控件的位置和关系的,有些类似于 RelativeLayout(相对布局管理器),但远比 RelativeLayout 强大。

例如,用户可以声明以下布局,如图 2.8 所示。

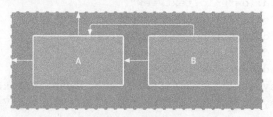

图 2.8　布局图示

- 视图 A 距离父布局顶部 16 dp。
- 视图 A 距离父布局左侧 16 dp。
- 视图 B 距离视图 A 右侧 16 dp。
- 视图 B 与视图 A 顶部对齐。

【例 2.4】　ConstraintLayout(约束布局管理器)实例

这个例子将使用 ConstraintLayout 创建一个简单的 Android 应用,下面将通过表 2.4 中的步骤来实现。

表 2.4　ConstraintLayout 实例步骤

步　骤	描　述
1	打开 Android Studio,创建一个 Android 应用,选择"Empty Activity"选项,在"Name"文本框中输入项目名,在"Package name"文本框中输入包名
2	在"Minimum SDK"下拉列表中选择"API 18: Android 4.3(Jelly Bean)"选项或其他 SDK
3	在"Language"下拉列表中选择"Java"选项
4	选择保存路径,单击"Finish"按钮
5	启动 Genymotion 模拟器,然后在 Android 工程项目中进行如下代码修改

修改工程项目的 res/layout 目录下的布局文件 activity_main.xml,修改代码如下:

```
<?xml version="1.0" encoding="utf-8"?>
<androidx.constraintlayout.widget.ConstraintLayout
    xmlns:android="http://schemas.android.com/apk/res/android"
    xmlns:app="http://schemas.android.com/apk/res-auto"
    xmlns:tools="http://schemas.android.com/tools"
    android:layout_width="match_parent"
    android:layout_height="match_parent"
```

```
            tools:context=".MainActivity">
    <TextView
        android:id="@+id/textView2"
        android:layout_width="wrap_content"
        android:layout_height="wrap_content"
        android:text="Hello World!"
        app:layout_constraintBottom_toBottomOf="parent"
        app:layout_constraintHorizontal_bias="0.205"
        app:layout_constraintLeft_toLeftOf="parent"
        app:layout_constraintRight_toRightOf="parent"
        app:layout_constraintTop_toTopOf="parent"
        app:layout_constraintVertical_bias="0.081" />
    <TextView
        android:id="@+id/textView"
        android:layout_width="wrap_content"
        android:layout_height="wrap_content"
        android:layout_marginStart="40dp"
        android:text="TextView"
        app:layout_constraintStart_toEndOf="@+id/textView2"
        app:layout_constraintTop_toTopOf="@+id/textView2" />
</androidx.constraintlayout.widget.ConstraintLayout>
```

程序在 Genymotion 中的运行效果如图 2.9 所示。

图 2.9　ConstraintLayout 实例运行效果

【例 2.5】　Space（空隙布局管理器）实例

Space 是一个轻量级的 View 子类，通常用于创建控件之间的间隙。接下来修改例 2.3，使用 Space、TableLayout 和 TableRow 创建一个简单的 Android 应用，下面将通过表 2.5 中的步骤来实现。

表 2.5　Space 实例步骤

步　骤	描　述
1	打开 Android Studio，创建一个 Android 应用，选择"Empty Activity"选项，在"Name"文本框中输入项目名，在"Package name"文本框中输入包名
2	在"Minimum SDK"下拉列表中选择"API 18: Android 4.3（Jelly Bean）"选项或其他 SDK
3	在"Language"下拉列表中选择"Java"选项
4	选择保存路径，单击"Finish"按钮
5	启动 Genymotion 模拟器，然后在 Android 工程项目中进行如下代码修改

修改工程项目的 res/layout 目录下的布局文件 activity_main.xml，修改代码如下：

```
<?xml version="1.0" encoding="utf-8"?>
<LinearLayout xmlns:android="http://schemas.android.com/apk/res/android"
    android:layout_width="fill_parent"
```

```xml
    android:layout_height="fill_parent"
    android:orientation="vertical" >

    <Space
        android:layout_width="match_parent"
        android:layout_height="250dp" />

<TableLayout
android:layout_width="fill_parent"
android:layout_height="fill_parent">

<TableRow
    android:layout_width="fill_parent"
    android:layout_height="fill_parent">
    <TextView
        android:text="Time"
        android:layout_width="wrap_content"
        android:layout_height="wrap_content"
        android:layout_column="1" />
    <TextClock
        android:layout_width="wrap_content"
        android:layout_height="wrap_content"
        android:id="@+id/textClock"
        android:layout_column="2" />
</TableRow>

<TableRow>
    <TextView
        android:text="First Name"
        android:layout_width="wrap_content"
        android:layout_height="wrap_content"
        android:layout_column="1" />
    <EditText
        android:width="200px"
        android:layout_width="wrap_content"
        android:layout_height="wrap_content" />
</TableRow>

<TableRow>
    <TextView
        android:text="Last Name"
        android:layout_width="wrap_content"
        android:layout_height="wrap_content"
        android:layout_column="1" />
    <EditText
        android:width="100px"
        android:layout_width="wrap_content"
        android:layout_height="wrap_content" />
</TableRow>
```

```xml
<TableRow
    android:layout_width="fill_parent"
    android:layout_height="fill_parent">
    <RatingBar
        android:layout_width="wrap_content"
        android:layout_height="wrap_content"
        android:id="@+id/ratingBar"
        android:layout_column="2" />
</TableRow>

<TableRow
    android:layout_width="fill_parent"
    android:layout_height="fill_parent" />
<TableRow
    android:layout_width="fill_parent"
    android:layout_height="fill_parent">
    <Button
        android:layout_width="wrap_content"
        android:layout_height="wrap_content"
        android:text="SUBMIT"
        android:id="@+id/button"
        android:layout_column="2" />
</TableRow>
</TableLayout>
</LinearLayout>
```

Space 界面设计效果如图 2.10 所示。

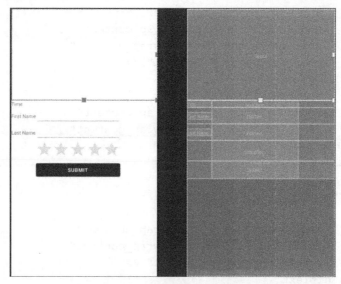

图 2.10　Space 界面设计效果

2.2 ListView（列表视图）

ListView 是 Android 中常用的 UI 控件之一，它将以垂直列表的形式列出要显示的列表项目。一般来说，ListView 都是和 Adapter（适配器）一起配合使用的，数组数据或者数据库数据都将通过 Adapter 把值传递给 ListView。ListView 实例运行效果如图 2.11 所示。

图 2.11　ListView 实例运行效果

Adapter 实际上是 UI 控件和数据源之间的一座桥梁，Adapter 从数据源中得到数据后将其传递给 AdapterView（适配器视图），AdapterView 则将数据呈现在 Spinner（列表选择框）、ListView（列表视图）、GridView（网格视图）等 UI 控件中。实际上，ListView、GridView 是 AdapterView 类的子类，它们通常和一个 Adapter 一起使用，并由 Adapter 负责收集外部数据，然后把数据项以 View 的形式显示在界面上。

Android 提供了几种类型的 Adapter，它们都是 Adapter 类的子类，常用的 Adapter 类型有 ArrayAdapter（数组适配器）、BaseAdapter（基本适配器）、CursorAdapter（游标适配器）、SimpleCursorAdapter（简单游标适配器）、SpinnerAdapter（列表适配器）和 WrapperListAdapter（封装列表适配器）。不同的适配器将用于绑定不同类型的数据，比如，ArrayAdapter 适用于绑定数组数据。

2.2.1　ListView（列表视图）的常用属性

ListView 的常用属性如表 2.6 所示。关于该控件的其他属性，可以参阅 Android 官方提供的完整的 API 文档。

表 2.6　ListView 的常用属性

属　　性	描　　述
android:id	用于设置 ListView 的名称
android:divider	用于为 ListView 设置分割条，既可以用颜色分割，也可以用 Drawable 资源分割

属 性	描 述
android:dividerHeight	用于设置分割条的高度
android:entries	用于通过数组资源为 ListView 指定列表项
android:footerDividersEnabled	用于设置是否在 footer view 之前绘制分割条，默认为 true，当设置为 false 时，表示不绘制。需要通过 ListView 提供的 addFooterView()方法为 ListView 添加 footer view
android:headerDividersEnabled	用于设置是否在 header view 之前绘制分割条，默认为 true，当设置为 false 时，表示不绘制。需要通过 ListView 提供的 addHeaderView()方法为 ListView 添加 header view

2.2.2 ArrayAdapter（数组适配器）

当需要为一个控件赋值，而赋值的数据源是一个数组时，就应该考虑使用 ArrayAdapter。首先创建一个适配器对象，然后使用 ArrayAdapter 类的构造方法 ArrayAdapter（Context context, int textViewResourceId, T[] objects）实例化一个 ArrayAdapter 对象。具体代码如下：

```
string[] StringArray=new String[]{"学生", "老师", "工作人员",}
ArrayAdapter adapter = new
ArrayAdapter<String>(this,R.layout.ListView,StringArray);
```

当创建好一个 ArrayAdapter 对象后，就可以简单地通过 ListView 的 setAdapter()方法进行调用了，代码如下：

```
ListView listView = (ListView) findViewById(R.id.listview);
ListView.setAdapter(adapter);
```

【例 2.6】 ListView 实例

这个例子将使用 ListView 创建一个简单的 Android 应用，下面将通过表 2.7 中的步骤来实现。

表 2.7 ListView 实例步骤

步 骤	描 述
1	打开 Android Studio，创建一个 Android 应用，选择"Empty Activity"选项，在"Name"文本框中输入项目名，在"Package name"文本框中输入包名
2	在"Minimum SDK"下拉列表中选择"API 18: Android 4.3（Jelly Bean）"选项或其他 SDK
3	在"Language"下拉列表中选择"Java"选项
4	选择保存路径，单击"Finish"按钮
5	在工程项目中找到 ras/layout 目录中的 activity_main.xml 文件，并在其中添加一个 ListView 控件
6	在 ras/layout 目录下添加一个名称为 activity_listview.xml 的文件。这个 XML 文件将用于显示所有的列表项，可以用这个文件对字体、间距、颜色等进行自定义
7	启动 Genymotion 模拟器，然后在 Android 工程项目中进行如下代码修改

修改工程项目的 res/layout 目录下的布局文件 activity_main.xml，修改代码如下：

```xml
<?xml version="1.0" encoding="utf-8"?>
<LinearLayout xmlns:android="http://schemas.android.com/apk/res/android"
    xmlns:tools="http://schemas.android.com/tools"
    android:layout_width="match_parent"
    android:layout_height="match_parent"
    android:orientation="vertical"
```

```xml
        tools:context=".ListActivity" >

    <ListView
        android:id="@+id/mobile_list"
        android:layout_width="match_parent"
        android:layout_height="wrap_content" >
    </ListView>

</LinearLayout>
```

修改新建工程项目的 res/layout 目录下的布局文件 activity_listview.xml，将默认添加的布局代码删除，然后添加如下代码：

```xml
<?xml version="1.0" encoding="utf-8"?>

<!-- Single List Item Design -->
<TextView xmlns:android="http://schemas.android.com/apk/res/android"
    android:id="@+id/label"
    android:layout_width="fill_parent"
    android:layout_height="fill_parent"
    android:padding="10dip"
    android:textSize="16dip"
    android:textStyle="bold" >
</TextView>
```

修改新建工程项目的 java/com.example.listdisplay 目录下的 Java 文件 MainActivity.java，将默认添加的布局代码删除，然后添加如下代码：

```java
package com.example.listdisplay;
import android.os.Bundle;
import android.support.v7.app.AppCompatActivity;
import android.widget.ArrayAdapter;
import android.widget.ListView;

public class MainActivity extends AppCompatActivity {
    // Array of strings...
    String[] mobileArray={"Android","IPhone","WindowsMobile","Blackberry",
                        "WebOS","Ubuntu","Windows7","Max OS X"};
    @Override
    protected void onCreate(Bundle savedInstanceState) {
        super.onCreate(savedInstanceState);
        setContentView(R.layout.activity_main);
        ArrayAdapter adapter = new ArrayAdapter<String>(this, R.layout.
                        activity_listview, mobileArray);

        ListView listView = (ListView) findViewById(R.id.mobile_list);
        listView.setAdapter(adapter);

    }
}
```

程序在 Genymotion 中的运行效果如图 2.12 所示。

图 2.12　ListView 实例运行效果

2.3　GridView（网格视图）

GridView 也是 Android 中常用的 UI 控件之一，它是按照行、列分布的方式来显示多个控件的，通常用于显示图片或图标等。GridView 实例运行效果如图 2.13 所示。

图 2.13　GridView 实例运行效果

GridView 的常用属性如表 2.8 所示。关于该控件的其他属性，可以参阅 Android 官方提供的完整的 API 文档。

表 2.8 GridView 的常用属性

属 性	描 述
android:id	用于设置 GridView 的名称
android:columnWidth	用于设置列的宽度
android:gravity	用于设置对齐方式
android:horizontalSpacing	用于设置各元素之间的水平间距
android:numColumns	用于设置列数,其值通常大于 1,如果其值较大,如 100,或者其值为 auto_fit,则将在可用空间中显示尽量多的列数
android:stretchMode	用于设置拉伸模式,包括以下几种类型。 • none:不拉伸 • spacingWidth:仅拉伸元素之间的间距 • columnWidth:仅拉伸表格元素本身 • spacingWidthUniform:将表格元素本身、元素之间的间距一起拉伸
android:verticalSpacing	用于设置各元素之间的垂直间距

GridView 和 ListView 类似,都需要通过 Adapter 来提供要显示的数据。在使用 GridView 时,通常使用 ImageAdapter(图片适配器)为 GridView 提供数据。下面通过一个具体的实例演示如何通过 ImageAdapter 指定内容的方式来创建 GridView。

【例 2.7】 GridView 实例

这个例子将使用 GridView 创建一个简单的 Android 应用,下面将通过表 2.9 中的步骤来实现。

注意: Android 虚拟机不允许单个程序中的位图占用超过 8MB 的内存。当加载大量图片时很容易出现内存溢出报错,而且如果位图突然要占用大量的内存,那么即使它和之前已经剩下的内存加起来并没有超过 8MB,系统也会报错,原因是它变"敏感"了。处理方法包括压缩图片大小、及时回收内存、使用缓存等。在例 2.7 中可以简单地压缩图片大小,然后将其复制到工程项目中,以避免程序报错。

表 2.9 GridView 实例步骤

步 骤	描 述
1	打开 Android Studio,创建一个 Android 应用,选择"Empty Activity"选项,在"Name"文本框中输入项目名,在"Package name"文本框中输入包名
2	在"Minimum SDK"下拉列表中选择"API 18: Android 4.3(Jelly Bean)"选项或其他 SDK
3	在"Language"下拉列表中选择"Java"选项
4	选择保存路径,单击"Finish"按钮
5	在工程项目中找到 ras/layout 目录中的 activity_main.xml 文件,并在其中添加一个 GridView 控件
6	将图片文件 sample0.jpg、sample1.jpg、sample2.jpg、sample3.jpg、sample4.jpg、sample5.jpg、sample6.jpg 和 sample7.jpg 复制、粘贴到工程项目的 res/drawable 目录下
7	在 java/com.example.griddisplay 目录下创建一个继承自 BaseAdapter(基本适配器)类的 ImageAdapter 子类文件,并将其命名为 ImageAdapter.java,用于为 GridView 控件提供图片数据
8	启动 Genymotion 模拟器,然后在 Android 工程项目中进行如下代码修改

修改工程项目的 res/layout 目录下的布局文件 activity_main.xml,修改代码如下:

```
<?xml version="1.0" encoding="utf-8"?>
<GridView xmlns:android="http://schemas.android.com/apk/res/android"
```

```
        android:id="@+id/gridview"
        android:layout_width="fill_parent"
        android:layout_height="fill_parent"
        android:columnWidth="90dp"
        android:numColumns="auto_fit"
        android:verticalSpacing="10dp"
        android:horizontalSpacing="10dp"
        android:stretchMode="columnWidth"
        android:gravity="center"
    />
```

修改工程项目的 java/com.example.griddisplay 目录下的 Java 文件 ImageAdapter.java，修改代码如下：

```
        package com.example.griddisplay;

        import android.content.Context;
        import android.view.View;
        import android.view.ViewGroup;
        import android.widget.BaseAdapter;
        import android.widget.GridView;
        import android.widget.ImageView;

        public class ImageAdapter extends BaseAdapter {
            private Context mContext;

            // Constructor
            public ImageAdapter(Context c) {
                mContext = c;
            }

            public int getCount() {
                return mThumbIds.length;
            }

            public Object getItem(int position) {
                return null;
            }

            public long getItemId(int position) {
                return 0;
            }

            // create a new ImageView for each item referenced by the Adapter
            public View getView(int position, View convertView, ViewGroup parent) {
                ImageView imageView;

                if (convertView == null) {
```

```
            imageView = new ImageView(mContext);
            imageView.setLayoutParams(new GridView.LayoutParams(85, 85));
            imageView.setScaleType(ImageView.ScaleType.CENTER_CROP);
            imageView.setPadding(8, 8, 8, 8);
        }
        else
        {
            imageView = (ImageView) convertView;
        }
        imageView.setImageResource(mThumbIds[position]);
        return imageView;
    }

    // Keep all Images in array
    public Integer[] mThumbIds = {
            R.drawable.sample2, R.drawable.sample3,
            R.drawable.sample4, R.drawable.sample5,
            R.drawable.sample6, R.drawable.sample7,
            R.drawable.sample0, R.drawable.sample1,
            R.drawable.sample2, R.drawable.sample3,
            R.drawable.sample4, R.drawable.sample5,
            R.drawable.sample6, R.drawable.sample7,
            R.drawable.sample0, R.drawable.sample1,
            R.drawable.sample2, R.drawable.sample3,
            R.drawable.sample4, R.drawable.sample5,
            R.drawable.sample6, R.drawable.sample7
    };
}
```

注意：在推出 Android 5.0 以后，它提供了很多新功能，也更新了 support v7，出现了 AppCompatActivity。AppCompatActivity 是用来替代 ActionBarActivity 的，在本书中所有的 AppCompatActivity 都可以被手动改成 Activity，不会影响程序运行。

修改工程项目的 java/com.example.griddisplay 目录下的 Java 文件 MainActivity.java，修改代码如下：

```java
package com.example.griddisplay;

import android.os.Bundle;
import android.support.v7.app.AppCompatActivity;
import android.widget.GridView;

public class MainActivity extends AppCompatActivity {

    @Override
    protected void onCreate(Bundle savedInstanceState) {
        super.onCreate(savedInstanceState);
        setContentView(R.layout.activity_main);
        GridView gridview = (GridView) findViewById(R.id.gridview);
```

```
        gridview.setAdapter(new ImageAdapter(this));
    }
}
```

程序在 Genymotion 中的运行效果如图 2.14 所示。

图 2.14　GridView 实例运行效果

【例 2.8】　GridView 实例扩展

下面对例 2.7 进行扩展：当单击网格视图中任意一个小图片时，将打开一个新页面并满屏显示这张图片。为了实现这一效果，需要新建一个 Activity（注意：所有的 Activity 都需要在 manifests/AndroidManifest.xml 文件中进行注册）。下面将通过表 2.10 中的步骤修改例 2.7 来实现。

表 2.10　GridView 扩展实例步骤

步骤	描　　述
1	打开 Android Studio，选择"Open an existing Android Studio project"（打开现有 Android Studio 工程）选项，打开例 2.7 的现有实例
2	在 java/com.example.griddisplay 目录下新建一个 Activity 类文件并将其命名为 SingleViewActivity.java
3	在 res/layout 目录下新建一个 XML 布局文件并将其命名为 single_view.xml
4	在 manifests 目录下的 AndroidManifest.xml 文件中为新建的 Activity 进行注册
5	启动 Genymotion 模拟器，然后在 Android 工程项目中进行如下代码修改

修改工程项目的 res/layout 目录下新建的布局文件 single_view.xml，修改代码如下：

```
<?xml version="1.0" encoding="utf-8"?>
<LinearLayout xmlns:android="http://schemas.android.com/apk/res/android"
    android:layout_width="match_parent"
    android:layout_height="match_parent"
    android:orientation="vertical">
```

```xml
<ImageView android:id="@+id/SingleView"
    android:layout_width="fill_parent"
    android:layout_height="fill_parent" />
```

```
</LinearLayout>
```

修改工程项目的java/com.example.griddisplay目录下新建的Activity类文件SingleViewActivity.java，修改代码如下：

```java
package com.example.griddisplay;

import android.app.Activity;
import android.content.Intent;
import android.os.Bundle;
import android.widget.ImageView;

public class SingleViewActivity extends Activity {
    @Override
    public void onCreate(Bundle savedInstanceState) {
        super.onCreate(savedInstanceState);
        setContentView(R.layout.single_view);

        // Get intent data
        Intent i = getIntent();

        // Selected image id
        int position = i.getExtras().getInt("id");
        ImageAdapter imageAdapter = new ImageAdapter(this);

        ImageView imageView = (ImageView) findViewById(R.id.SingleView);
        imageView.setImageResource(imageAdapter.mThumbIds[position]);
    }
}
```

修改工程项目的java/com.example.griddisplay目录下的Java文件MainActivity.java，修改代码如下：

```java
package com.example.griddisplay;

import android.content.Intent;
import android.os.Bundle;
import android.support.v7.app.AppCompatActivity;
import android.view.View;
import android.widget.AdapterView;
import android.widget.AdapterView.OnItemClickListener;
import android.widget.GridView;

public class MainActivity extends AppCompatActivity {
```

```java
@Override
protected void onCreate(Bundle savedInstanceState) {
    super.onCreate(savedInstanceState);
    setContentView(R.layout.activity_main);
    GridView gridview = (GridView) findViewById(R.id.gridview);
    gridview.setAdapter(new ImageAdapter(this));

    gridview.setOnItemClickListener(new OnItemClickListener() {
        public void onItemClick(AdapterView<?> parent, View v, int
            position, long id) {
            // Send intent to SingleViewActivity
            Intent i = new Intent(getApplicationContext(), SingleViewActivity.
                class);

            // Pass image index
            i.putExtra("id", position);
            startActivity(i);
        }
    });
}
```

修改工程项目的 manifests 目录下的工程配置文件 AndroidManifest.xml，修改代码如下：

```xml
<?xml version="1.0" encoding="utf-8"?>
<manifest xmlns:android="http://schemas.android.com/apk/res/android"
    package="com.example.griddisplay" >

    <application
        android:allowBackup="true"
        android:icon="@mipmap/ic_launcher"
        android:label="@string/app_name"
        android:supportsRtl="true"
        android:theme="@style/AppTheme" >
        <activity android:name=".MainActivity" >
            <intent-filter>
                <action android:name="android.intent.action.MAIN" />

                <category android:name="android.intent.category.LAUNCHER" />
            </intent-filter>
        </activity>
        <activity android:name=".SingleViewActivity"></activity>
    </application>

</manifest>
```

程序在 Genymotion 中的运行效果如图 2.15 和图 2.16 所示。

图 2.15　GridView 实例扩展运行效果（1）

图 2.16　GridView 实例扩展运行效果（2）

2.4　Android UI 控件

Android 应用程序的人机交互界面由很多 Android 组件组成。Android 提供了很多 UI 控件，如图 2.17 所示，部分常用 UI 控件包括 TextView（文本框）、EditText（编辑框）、AutoCompleteTextView（自动填充文本框）、Button（普通按钮）、ImageButton（图片按钮）、CheckBox（复选框）、ToggleButton（开关按钮）、RadioButton（单选按钮）等。

2.4.1　TextView（文本框）

在 Android 中，TextView 用于在屏幕中显示文本。TextView 是一个简单的 View，因此它可以直接继承 View 对象。

图 2.17　部分常用 UI 控件

TextView 的常用属性如表 2.11 所示，关于该控件的其他属性，可以参阅 Android 官方提供的完整的 API 文档。

表 2.11　TextView 的常用属性

属　性	描　述
android:id	用于设置文本框的名称
android:capitalize	用于设置输入英文字母时是否自动转变为大写字母 ● 0 表示不自动转变为大写字母 ● 1 表示一句话的首字母自动转变为大写字母 ● 2 表示每个单词的首字母自动转变为大写字母 ● 3 表示所有字母自动转变为大写字母
android:cursorVisible	当进行编辑时，是否显示光标 ● true 表示显示光标 ● false 表示不显示光标（默认值）
android:editable	如果设置为 true，则表示文本框可以进行编辑
android:fontFamily	用于设置文本框的字体样式

续表

属性	描述
android:gravity	用于设置文本框的对齐方式，可选值有 top、left、right、center_vertical、fill_vertical、center_horizontal、fill_horizontal、center、fill、clip_vertical 和 clip_horizontal 等。也可以同时设定这些属性值，并将各属性值用"|"（竖线）隔开。例如，要指定控件在右下角对齐，则可以使用属性值 right|bottom
android:hint	用于设置当文本框中文本内容为空时，默认显示的提示文本
android:inputType	用于指定当前文本框显示内容的文本类型，其可选值有 textpassword、textEmailAddress、phone 和 date 等，可以同时指定多个，并使用"|"（竖线）隔开
android:maxHeight	用于设置文本框的最大高度
android:maxWidth	用于设置文本框的最大宽度
android:minHeight	用于设置文本框的最小高度
android:minWidth	用于设置文本框的最小宽度
android:password	用于设置当文本框可以编辑时，里面的文本是否显示为隐藏密码样式"."，true 表示显示，false 表示不显示
android:phoneNumber	用于设置当文本框可以编辑时，里面的文本是否显示为电话号码样式，true 表示显示，false 表示不显示
android:text	用于指定文本框中显示的文本内容
android:textAllCaps	用于设置文本框中的英文字母是否全部转变为大写字母，true 表示转变为大写字母，false 表示不转变为大写字母
android:textColor	用于设置文本框内文本的颜色，其属性值可以是"#rgb"、"#argb"、"#rrggbb"或 "#aarrggbb"
android:textColorHighlight	用于设置文本框内文本的颜色是否高亮
android:textColorHint	用于设置文本框提示信息的颜色，其属性值可以是"#rgb"、"#argb"、"#rrggbb"或 "#aarrggbb"
android:textIsSelectable	用于设置当文本框不可以编辑时，文本框中的文字是否可以被选中，true 表示可以，false 表示不可以
android:textSize	用于设置文本框中文本的字号大小，其属性由代表大小的数值和单位组成，其单位可以是 px、pt、sp 和 in 等
android:textStyle	用于设置文本框中文本的样式，可以同时指定多个，使用"|"（竖线）隔开 • 0 表示正常 • 1 表示加粗 • 2 表示斜体

【例 2.9】 TextView 实例

这个例子将使用 TextView 来创建一个简单的 Android 应用，下面将通过表 2.12 中的步骤来实现。

表 2.12 TextView 实例步骤

步骤	描述
1	打开 Android Studio，创建一个 Android 应用，选择"Empty Activity"选项，在"Name"文本框中输入项目名，在"Package name"文本框中输入包名
2	在"Minimum SDK"下拉列表中选择"API 18: Android 4.3（Jelly Bean）"选项或其他 SDK
3	在"Language"下拉列表中选择"Java"选项
4	选择保存路径，单击"Finish"按钮
5	在工程项目中找到 ras/layout 目录中的 activity_main.xml 文件，并在其中添加一个 TextView 控件
6	启动 Genymotion 模拟器，然后在 Android 工程项目中进行如下代码修改

修改新建工程项目的 res/layout 目录下的布局文件 activity_main.xml，修改代码如下：

```xml
<?xml version="1.0" encoding="utf-8"?>
<RelativeLayout xmlns:android="http://schemas.android.com/apk/res/android"
    xmlns:tools="http://schemas.android.com/tools"
    android:layout_width="match_parent"
    android:layout_height="match_parent"
    android:paddingLeft="@dimen/activity_horizontal_margin"
    android:paddingRight="@dimen/activity_horizontal_margin"
    android:paddingTop="@dimen/activity_vertical_margin"
    android:paddingBottom="@dimen/activity_vertical_margin"
    tools:context=".MainActivity">

    <TextView
        android:id="@+id/text_id"
        android:layout_width="300dp"
        android:layout_height="200dp"
        android:capitalize="characters"
        android:text="hello_world"
        android:textColor="@android:color/holo_blue_dark"
        android:textColorHighlight="@android:color/primary_text_dark"
        android:layout_centerVertical="true"
        android:layout_alignParentEnd="true"
        android:textSize="50dp" />
</RelativeLayout>
```

程序在 Genymotion 中的运行效果如图 2.18 所示。

图 2.18　TextView 实例运行效果

2.4.2　EditText（编辑框）

　　EditText 用于在屏幕中显示文本输入框。在 Android 的 EditText 中，可以输入单行文本，也可以输入多行文本，还可以输入指定格式的文本（如密码、电话号码、E-mail 地址等）。

下面介绍 EditText 的常用属性，关于该控件的其他属性，可以参阅 Android 官方提供的完整的 API 文档。

（1）继承于 android.widget.TextView 类的 EditText 的常用属性如表 2.13 所示。

表 2.13　继承于 android.widget.TextView 类的 EditText 的常用属性

属　　性	描　　述
android:autoText	若被选中，则表示对编辑框中的文本自动更正拼写错误
android:drawableBottom	用于在编辑框内文本的底端绘制指定图片，该图片可以是放在 res/drawable 目录下的图片，通过"@drawable/文件名（不包括文件的扩展名）"设置
android:drawableRight	用于在编辑框内文本的右侧绘制指定图片，该图片可以是放在 res/drawable 目录下的图片，通过"@drawable/文件名（不包括文件的扩展名）"设置
android:editable	用于设置编辑框是否能被编辑，默认值为 true
android:text	用于设置编辑框中默认显示的文本

（2）继承于 android.view.View 类的 EditText 的常用属性如表 2.14 所示。

表 2.14　继承于 android.view.View 类的 EditText 的常用属性

属　　性	描　　述
android:background	用于设置编辑框的背景图片，该图片可以是放在 res/drawable 目录下的图片，通过"@drawable/文件名（不包括文件的扩展名）"设置
android:contentDescription	用于设置编辑框的简单描述文字
android:id	用于设置编辑框的名称
android:onClick	用于设置单击事件响应方法的方法名
android:visibility	用于设置编辑框是否可见

【例 2.10】 EditText 实例

这个例子将使用 EditText 来创建一个简单的 Android 应用，下面将通过表 2.15 中的步骤来实现。

表 2.15　EditText 实例步骤

步　骤	描　　述
1	打开 Android Studio，创建一个 Android 应用，选择"Empty Activity"选项，在"Name"文本框中输入项目名，在"Package name"文本框中输入包名
2	在"Minimum SDK"下拉列表中选择"API 18: Android 4.3（Jelly Bean）"选项或其他 SDK
3	在"Language"下拉列表中选择"Java"选项
4	选择保存路径，单击"Finish"按钮
5	在工程项目中找到 ras/layout 目录中的 activity_main.xml 文件，并在其中添加一个 EditText 控件
6	启动 Genymotion 模拟器，然后在 Android 工程项目中进行如下代码修改

修改新建工程项目的 res/layout 目录下的布局文件 activity_main.xml，修改代码如下：

```
<?xml version="1.0" encoding="utf-8"?>
<RelativeLayout xmlns:android="http://schemas.android.com/apk/res/android"
    xmlns:tools="http://schemas.android.com/tools"
    android:layout_width="match_parent"
    android:layout_height="match_parent"
    android:paddingLeft="@dimen/activity_horizontal_margin"
```

```xml
        android:paddingRight="@dimen/activity_horizontal_margin"
        android:paddingTop="@dimen/activity_vertical_margin"
        android:paddingBottom="@dimen/activity_vertical_margin"
        tools:context=".MainActivity">

    <TextView
        android:id="@+id/textView1"
        android:layout_width="wrap_content"
        android:layout_height="wrap_content"
        android:layout_alignParentLeft="true"
        android:layout_alignParentTop="true"
        android:layout_marginLeft="14dp"
        android:layout_marginTop="18dp"
        android:text="@string/example_edittext" />

    <Button
        android:id="@+id/button"
        android:layout_width="wrap_content"
        android:layout_height="wrap_content"
        android:layout_alignLeft="@+id/textView1"
        android:layout_below="@+id/textView1"
        android:layout_marginTop="130dp"
        android:text="@string/show_the_text" />

    <EditText
        android:id="@+id/edittext"
        android:layout_width="fill_parent"
        android:layout_height="wrap_content"
        android:layout_alignLeft="@+id/button"
        android:layout_below="@+id/textView1"
        android:layout_marginTop="61dp"
        android:ems="10"
        android:text="@string/enter_text" android:inputType="text" />
</RelativeLayout>
```

修改新建工程项目的 values 目录下的字符串文件 strings.xml，修改代码如下：

```xml
<?xml version="1.0" encoding="utf-8"?>
<resources>
    <string name="app_name">demo</string>
    <string name="action_settings">Settings</string>
    <string name="example_edittext">Example showing EditText</string>
    <string name="show_the_text">SHOW THE TEXT</string>
    <string name="enter_text">text changes</string>
</resources>
```

程序在 Genymotion 中的运行效果如图 2.19 所示。

图 2.19　EditText 实例运行效果

2.4.3　AutoCompleteTextView（自动填充文本框）

AutoCompleteTextView 用于实现当用户输入一定的字符后，显示一个下拉列表，供用户选择其中的选项，当用户选择某个选项后，就按用户的选择自动填写文本框。

AutoCompleteTextView 的常用属性如表 2.16 所示，关于该控件的其他属性，可以参阅 Android 官方提供的完整的 API 文档。

表 2.16　AutoCompleteTextView 的常用属性

属　　性	描　　述
android:completionHint	用于为弹出的下拉列表指定提示标题
android:completionHintView	用于设置弹出的下拉列表底部信息的样式
android:completionThreshold	用于指定用户至少输入几个字符才会显示提示
android:dropDownAnchor	它的值是一个 View 的 ID，在被指定后，AutoCompleteTextView 会在这个 View 下弹出自动提示
android:dropDownHeight	用于指定下拉列表的高度
android:dropDownHorizontalOffset	用于指定下拉列表与文本之间的水平偏移。下拉列表默认与文本框左对齐
android:dropDownSelector	用于设置下拉列表的显示效果
android:dropDownVerticalOffset	用于设置下拉列表与文本之间的垂直偏移。下拉列表默认紧跟文本框
android:dropDownWidth	用于设置下拉列表的宽度
android:popupBackground	用于设置下拉列表的背景

【例 2.11】　AutoCompleteTextView 实例

这个例子将使用 AutoCompleteTextView 来创建一个简单的 Android 应用，下面将通过表 2.17 中的步骤来实现。

第 2 章 Android UI 设计

表 2.17 AutoCompleteTextView 实例步骤

步骤	描述
1	打开 Android Studio，创建一个 Android 应用，选择"Empty Activity"选项，在"Name"文本框中输入项目名，在"Package name"文本框中输入包名
2	在"Minimum SDK"下拉列表中选择"API 18: Android 4.3（Jelly Bean）"选项或其他 SDK
3	在"Language"下拉列表中选择"Java"选项
4	选择保存路径，单击"Finish"按钮
5	在工程项目中找到 ras/layout 目录中的 activity_main.xml 文件，并在其中添加一个 AutoCompleteTextView 控件
6	启动 Genymotion 模拟器，然后在 Android 工程项目中进行如下代码修改

修改新建工程项目的 res/layout 目录下的布局文件 activity_main.xml，修改代码如下：

```xml
<?xml version="1.0" encoding="utf-8"?>
<RelativeLayout xmlns:android="http://schemas.android.com/apk/res/android"
    xmlns:tools="http://schemas.android.com/tools"
    android:layout_width="match_parent"
    android:layout_height="match_parent"
    android:paddingLeft="@dimen/activity_horizontal_margin"
    android:paddingRight="@dimen/activity_horizontal_margin"
    android:paddingTop="@dimen/activity_vertical_margin"
    android:paddingBottom="@dimen/activity_vertical_margin"
    tools:context=".MainActivity">

    <TextView
        android:id="@+id/textView2"
        android:layout_width="wrap_content"
        android:layout_height="wrap_content"
        android:layout_alignParentTop="true"
        android:layout_centerHorizontal="true"
        android:layout_marginTop="25dp"
        android:text="Example showing AutoCompleteTextView" />

    <AutoCompleteTextView
        android:id="@+id/autoCompleteTextView1"
        android:layout_width="wrap_content"
        android:layout_height="wrap_content"
        android:layout_alignLeft="@+id/textView2"
        android:layout_below="@+id/textView2"
        android:layout_marginTop="54dp"
        android:ems="10" />
</RelativeLayout>
```

修改新建工程项目的 java/com.example.autocompletetextview 目录下的 Java 文件 MainActivity.java，修改代码如下：

```java
package com.example.autocompletetextview;

import android.os.Bundle;
import android.support.v7.app.AppCompatActivity;
```

```java
import android.widget.ArrayAdapter;
import android.widget.AutoCompleteTextView;

public class MainActivity extends AppCompatActivity {

    AutoCompleteTextView autocomplete;

    String[] arr = { "Paries,France", "PA,United States","Parana,Brazil",
                "Padua,Italy", "Pasadena,CA,United States"};

    @Override
    protected void onCreate(Bundle savedInstanceState) {
        super.onCreate(savedInstanceState);
        setContentView(R.layout.activity_main);

        autocomplete = (AutoCompleteTextView)
                findViewById(R.id.autoCompleteTextView1);

        ArrayAdapter<String> adapter = new ArrayAdapter<String>
                (this,android.R.layout.select_dialog_item, arr);

        autocomplete.setThreshold(2);
        autocomplete.setAdapter(adapter);
    }
}
```

在 AutoCompleteTextView 控件中输入 "pa"，自动提示下拉列表在 Genymotion 中的运行效果如图 2.20 所示。

图 2.20　AutoCompleteTextView 实例运行效果

2.4.4 Button（普通按钮）

Button 用于在 UI 上生成一个可以单击的按钮。当用户单击按钮时，将会触发一个 onClick 事件，可以通过为按钮添加单击事件监听器来指定所要触发的动作。

下面介绍 Button 的常用属性，关于该控件的其他属性，可以参阅 Android 官方提供的完整的 API 文档。

（1）继承于 android.widget.TextView 类的 Button 的常用属性如表 2.18 所示。

表 2.18　继承于 android.widget.TextView 类的 Button 的常用属性

属　性	描　述
android:autoText	若被选中，则表示对按钮上显示的文本自动更正拼写错误
android:drawableBottom	用于在按钮上显示文本的底端绘制指定图片，该图片可以是放在 res/drawable 目录下的图片，通过"@drawable/文件名（不包括文件的扩展名）"设置
android:drawableRight	用于在按钮上显示文本的右侧绘制指定图片，该图片可以是放在 res/drawable 目录下的图片，通过"@drawable/文件名（不包括文件的扩展名）"设置
android:enabled	用于设置按钮是否可用
android:text	用于设置按钮上显示的文字

（2）继承于 android.view.View 类的 Button 的常用属性如表 2.19 所示。

表 2.19　继承于 android.view.View 类的 Button 的常用属性

属　性	描　述
android:background	用于设置按钮的背景图片，该图片可以是放在 res/drawable 目录下的图片，通过"@drawable/文件名（不包括文件的扩展名）"设置
android:contentDescription	用于设置按钮的简单描述文字
android:id	用于设置按钮的名称
android:onClick	用于设置单击事件响应方法的方法名
android:visibility	用于设置按钮是否可见

【例 2.12】　Button 实例

这个例子将使用 Button 来创建一个简单的 Android 应用，下面将通过表 2.20 中的步骤来实现。

表 2.20　Button 实例步骤

步　骤	描　述
1	打开 Android Studio，创建一个 Android 应用，选择"Empty Activity"选项，在"Name"文本框中输入项目名，在"Package name"文本框中输入包名
2	在"Minimum SDK"下拉列表中选择"API 18: Android 4.3（Jelly Bean）"选项或其他 SDK
3	在"Language"下拉列表中选择"Java"选项
4	选择保存路径，单击"Finish"按钮
5	在工程项目中找到 ras/layout 目录中的 activity_main.xml 文件，并在其中添加一个 TextView、一个 ImageButton、一个 EditText 和一个 Button 控件，按如图 2.21 所示的运行效果进行摆放
6	将工程项目的 res/mipmap 目录下的图片文件 ic_launcher.png 复制、粘贴到 res/drawable 目录下
7	启动 Genymotion 模拟器，然后在 Android 工程项目中进行如下代码修改

修改新建工程项目的 res/layout 目录下的布局文件 activity_main.xml，修改代码如下：

```xml
<?xml version="1.0" encoding="utf-8"?>
<RelativeLayout xmlns:android="http://schemas.android.com/apk/res/android"
    xmlns:tools="http://schemas.android.com/tools"
    android:layout_width="match_parent"
    android:layout_height="match_parent"
    android:paddingLeft="@dimen/activity_horizontal_margin"
    android:paddingRight="@dimen/activity_horizontal_margin"
    android:paddingTop="@dimen/activity_vertical_margin"
    android:paddingBottom="@dimen/activity_vertical_margin"
    tools:context=".MainActivity">

    <TextView
        android:id="@+id/textView1"
        android:layout_width="wrap_content"
        android:layout_height="wrap_content"
        android:text="Button Control"
        android:layout_alignParentTop="true"
        android:layout_centerHorizontal="true"
        android:textSize="30dp" />

    <ImageButton
        android:layout_width="wrap_content"
        android:layout_height="wrap_content"
        android:id="@+id/imageButton"
        android:src="@drawable/ic_launcher"
        android:layout_centerHorizontal="true"
        android:layout_below="@+id/textView1" />

    <Button
        android:layout_width="wrap_content"
        android:layout_height="wrap_content"
        android:text="Button"
        android:id="@+id/button"
        android:layout_alignStart="@+id/imageButton"
        android:layout_below="@+id/imageButton"
        android:layout_alignRight="@+id/imageButton"
        android:layout_alignEnd="@+id/imageButton" />
</RelativeLayout>
```

修改新建工程项目的 java/com.example.button 目录下的 Java 文件 MainActivity.java,修改代码如下:

```java
package com.example.button;

import android.app.Activity;
import android.os.Bundle;
import android.view.View;
import android.widget.Button;
import android.widget.Toast;
```

```java
public class MainActivity extends Activity {
    Button b1;
    @Override
    protected void onCreate(Bundle savedInstanceState) {
        super.onCreate(savedInstanceState);
        setContentView(R.layout.activity_main);

        b1=(Button)findViewById(R.id.button);
        b1.setOnClickListener(new View.OnClickListener() {
            @Override
            public void onClick(View v) {
                Toast.makeText(MainActivity.this, "YOUR MESSAGE", Toast.
                        LENGTH_LONG).show();
            }
        });
    }
}
```

在 Genymotion 中运行程序，然后单击 Button 控件，其效果如图 2.21 所示。

图 2.21　Button 实例运行效果图

2.4.5　ImageButton（图片按钮）

图片按钮和普通按钮的使用方法基本相同，只不过图片按钮具有 android:src 属性，可以用来设置在按钮中显示的图片。与普通按钮一样，也可以通过为图片按钮添加单击事件监听器来指定所要触发的动作。

下面介绍 ImageButton 的常用属性，关于该控件的其他属性，可以参阅 Android 官方提供的完整的 API 文档。

（1）继承于 android.widget.ImageView 类的 ImageButton 的常用属性如表 2.21 所示。

表 2.21 继承于 android.widget.ImageView 类的 ImageButton 的常用属性

属性	描述
android:adjustViewBounds	如果设置为 true，则 ImageButton 将调整其边框以适应图片大小
android:baseline	用于设置 ImageButton 的基线
android:baselineAlignBottom	值为 true 表示图片的基线与按钮底部对齐
android:cropToPadding	值为 true 表示 ImageButton 会剪切图片以适应内边距的大小
android:src	用于设置要显示的图片资源位置，该图片可以是放在 res/drawable 目录下的图片，通过"@drawable/文件名（不包括文件的扩展名）"设置

（2）继承于 android.view.View 类的 ImageButton 的常用属性如表 2.22 所示。

表 2.22 继承于 android.view.View 类的 ImageButton 的常用属性

属性	描述
android:background	用于设置图片按钮的背景图片，该图片可以是放在 res/drawable 目录下的图片，通过"@drawable/文件名（不包括文件的扩展名）"设置
android:contentDescription	用于设置图片按钮的简单描述文字
android:id	用于设置按钮的名称
android:onClick	用于设置单击事件响应方法的方法名
android:visibility	用于设置图片按钮是否可见

【例 2.13】 ImageButton 实例

这个例子将使用 ImageButton 来创建一个简单的 Android 应用，下面将通过表 2.23 中的步骤来实现。

表 2.23 ImageButton 实例步骤

步骤	描述
1	打开 Android Studio，创建一个 Android 应用，选择 "Empty Activity" 选项，在 "Name" 文本框中输入项目名，在 "Package name" 文本框中输入包名
2	在 "Minimum SDK" 下拉列表中选择 "API 18: Android 4.3（Jelly Bean）" 选项或其他 SDK
3	在 "Language" 下拉列表中选择 "Java" 选项
4	选择保存路径，单击 "Finish" 按钮
5	在工程项目中找到 ras/layout 目录中的 activity_main.xml 文件，并在其中添加一个 TextView 和一个 ImageButton 控件，按如图 2.22 所示的运行效果进行摆放
6	将工程项目的 res/mipmap 目录下的图片文件 ic_launcher.png 复制、粘贴到 res/drawable 目录下
7	启动 Genymotion 模拟器，然后在 Android 工程项目中进行如下代码修改

修改新建工程项目的 res/layout 目录下的布局文件 activity_main.xml，修改代码如下：

```
<?xml version="1.0" encoding="utf-8"?>
<RelativeLayout xmlns:android="http://schemas.android.com/apk/res/android"
    xmlns:tools="http://schemas.android.com/tools"
    android:layout_width="match_parent"
    android:layout_height="match_parent"
    android:paddingLeft="@dimen/activity_horizontal_margin"
    android:paddingRight="@dimen/activity_horizontal_margin"
    android:paddingTop="@dimen/activity_vertical_margin"
    android:paddingBottom="@dimen/activity_vertical_margin"
```

```xml
    tools:context=".MainActivity">

    <TextView android:text="ImageButton"
        android:layout_width="wrap_content"
        android:layout_height="wrap_content"
        android:textSize="30dp"
        android:layout_alignParentTop="true"
        android:layout_alignRight="@+id/imageButton"
        android:layout_alignEnd="@+id/imageButton" />

    <ImageButton
        android:layout_width="wrap_content"
        android:layout_height="wrap_content"
        android:id="@+id/imageButton"
        android:layout_centerVertical="true"
        android:layout_centerHorizontal="true"
        android:src="@drawable/ic_launcher" />
</RelativeLayout>
```

修改新建工程项目的 java/com.example.imagebutton 目录下的 Java 文件 MainActivity.java，修改代码如下：

```java
package com.example.imagebutton;

import android.os.Bundle;
import android.support.v7.app.AppCompatActivity;
import android.view.View;
import android.widget.ImageButton;
import android.widget.Toast;

public class MainActivity extends AppCompatActivity {
    ImageButton imgButton;
    @Override
    protected void onCreate(Bundle savedInstanceState) {
        super.onCreate(savedInstanceState);
        setContentView(R.layout.activity_main);
        imgButton =(ImageButton)findViewById(R.id.imageButton);
        imgButton.setOnClickListener(new View.OnClickListener() {
            @Override
            public void onClick(View v) {
                Toast.makeText(getApplicationContext(), "You download is resumed", Toast.LENGTH_LONG).show();
            }
        });
    }
}
```

在 Genymotion 中运行程序，然后单击 ImageButton 控件，其效果如图 2.22 所示。

图 2.22　ImageButton 实例运行效果

2.4.6　CheckBox（复选框）

在 Android 中，单选按钮和复选框都继承于普通按钮，因此，它们都可以直接使用普通按钮支持的各种属性和方法，与普通按钮不同的是，它们提供了可选中的功能。

在默认情况下，CheckBox 显示为一个方框图标，并且在该图标旁边会放置一些说明文字。CheckBox 可以进行多选设置，每个复选框都提供"选中"和"不选中"两种状态。

下面介绍 CheckBox 的常用属性，关于该控件的其他属性，可以参阅 Android 官方提供的完整的 API 文档。

（1）继承于 android.widget.TextView 类的 CheckBox 的常用属性如表 2.24 所示。

表 2.24　继承于 android.widget.TextView 类的 CheckBox 的常用属性

属性	描述
android:autoText	若被选中，则表示对复选框的说明文本自动更正拼写错误
android:drawableBottom	用于在复选框的说明文本的底端绘制指定图片，该图片可以是放在 res/drawable 目录下的图片，通过"@drawable/文件名（不包括文件的扩展名）"设置
android:drawableRight	用于在复选框的说明文本的右侧绘制指定图片，该图片可以是放在 res/drawable 目录下的图片，通过"@drawable/文件名（不包括文件的扩展名）"设置
android:editable	用于设置复选框的说明文本是否能被编辑，默认值为 true
android:text	用于设置复选框的默认说明文本

（2）继承于 android.view.View 类的 CheckBox 的常用属性如表 2.25 所示。

表 2.25　继承于 android.view.View 类的 CheckBox 的常用属性

属性	描述
android:background	用于设置复选框的背景图片，该图片可以是放在 res/drawable 目录下的图片，通过"@drawable/文件名（不包括文件的扩展名）"设置
android:contentDescription	用于设置复选框的简单描述文字
android:id	用于设置复选框的名称

属性	描述
android:onClick	用于设置单击事件响应方法的方法名
android:visibility	用于设置复选框是否可见

【例 2.14】 CheckBox 实例

这个例子将使用 CheckBox 来创建一个简单的 Android 应用,下面将通过表 2.26 中的步骤来实现。

表 2.26 CheckBox 实例步骤

步骤	描述
1	打开 Android Studio,创建一个 Android 应用,选择"Empty Activity"选项,在"Name"文本框中输入项目名,在"Package name"文本框中输入包名
2	在"Minimum SDK"下拉列表中选择"API 18: Android 4.3(Jelly Bean)"选项或其他 SDK
3	在"Language"下拉列表中选择"Java"选项
4	选择保存路径,单击"Finish"按钮
5	在工程项目中找到 ras/layout 目录中的 activity_main.xml 文件,并在其中添加一个 TextView、两个 CheckBox、两个 Button 和一个 ImageButton 控件,按如图 2.23 所示的运行效果进行摆放
6	将工程项目的 res/mipmap 目录下的图片文件 ic_launcher.png 复制、粘贴到 res/drawable 目录下
7	启动 Genymotion 模拟器,然后在 Android 工程项目中进行如下代码修改

修改新建工程项目的 res/layout 目录下的布局文件 activity_main.xml,修改代码如下:

```xml
<?xml version="1.0" encoding="utf-8"?>
<RelativeLayout xmlns:android="http://schemas.android.com/apk/res/android"
    xmlns:tools="http://schemas.android.com/tools"
    android:layout_width="match_parent"
    android:layout_height="match_parent"
    android:paddingLeft="@dimen/activity_horizontal_margin"
    android:paddingRight="@dimen/activity_horizontal_margin"
    android:paddingTop="@dimen/activity_vertical_margin"
    android:paddingBottom="@dimen/activity_vertical_margin"
    tools:context=".MainActivity">
    <TextView
        android:id="@+id/textView1"
        android:layout_width="wrap_content"
        android:layout_height="wrap_content"
        android:text="Example of checkbox"
        android:layout_alignParentTop="true"
        android:layout_centerHorizontal="true"
        android:textSize="30dp" />

    <CheckBox
        android:id="@+id/checkBox1"
        android:layout_width="wrap_content"
        android:layout_height="wrap_content"
        android:text="Do you like Tutorials Point"
        android:layout_above="@+id/button"
        android:layout_centerHorizontal="true" />
```

```xml
<CheckBox
    android:id="@+id/checkBox2"
    android:layout_width="wrap_content"
    android:layout_height="wrap_content"
    android:text="Do you like android "
    android:checked="false"
    android:layout_above="@+id/checkBox1"
    android:layout_alignLeft="@+id/checkBox1"
    android:layout_alignStart="@+id/checkBox1" />

<Button
    android:layout_width="wrap_content"
    android:layout_height="wrap_content"
    android:text="OK"
    android:id="@+id/button"
    android:layout_alignParentBottom="true"
    android:layout_alignLeft="@+id/checkBox1"
    android:layout_alignStart="@+id/checkBox1" />

<Button
    android:layout_width="wrap_content"
    android:layout_height="wrap_content"
    android:text="CANCEL"
    android:id="@+id/button2"
    android:layout_alignParentBottom="true"
    android:layout_alignRight="@+id/textView1"
    android:layout_alignEnd="@+id/textView1" />

<ImageButton
    android:layout_width="wrap_content"
    android:layout_height="wrap_content"
    android:id="@+id/imageButton"
    android:src="@drawable/ic_launcher"
    android:layout_centerVertical="true"
    android:layout_centerHorizontal="true" />
</RelativeLayout>
```

修改新建工程项目的 java/com.example.checkbox 目录下的 Java 文件 MainActivity.java，修改代码如下：

```java
package com.example.checkbox;

import android.os.Bundle;
import android.support.v7.app.AppCompatActivity;
import android.view.View;
import android.widget.Button;
import android.widget.CheckBox;
import android.widget.Toast;

public class MainActivity extends AppCompatActivity {
    CheckBox ch1,ch2;
    Button b1,b2;
```

```java
@Override
protected void onCreate(Bundle savedInstanceState) {
    super.onCreate(savedInstanceState);
    setContentView(R.layout.activity_main);
    ch1=(CheckBox)findViewById(R.id.checkBox1);
    ch2=(CheckBox)findViewById(R.id.checkBox2);

    b1=(Button)findViewById(R.id.button);
    b2=(Button)findViewById(R.id.button2);
    b2.setOnClickListener(new View.OnClickListener() {
        @Override
        public void onClick(View v) {
            finish();
        }
    });

    b1.setOnClickListener(new View.OnClickListener() {
        @Override
        public void onClick(View v) {
            StringBuffer result = new StringBuffer();
            result.append("Thanks : ").append(ch1.isChecked());
            result.append("\nThanks: ").append(ch2.isChecked());
            Toast.makeText(MainActivity.this, result.toString(), Toast.
                    LENGTH_LONG).show();
        }
    });
}
```

在 Genymotion 中运行程序，然后选中其中一个 CheckBox 控件或者同时选中两个 CheckBox 控件，并单击"OK"按钮，效果如图 2.23 所示。

图 2.23 CheckBox 实例运行效果

2.4.7 ToggleButton（开关按钮）

ToggleButton 是 Android 中比较简单的一个控件，具有选中和未选中两种状态，并且需要为不同的状态设置不同的说明文本，同时具有一个显示开/关的指示灯。

ToggleButton 的常用属性如表 2.27 所示，关于该控件的其他属性，可以参阅 Android 官方提供的完整的 API 文档。

表 2.27 ToggleButton 的常用属性

属性	描述
android:disabledAlpha	用于设置开关按钮在禁用时的透明度
android:textOff	用于设置开关按钮未被选中时的说明文本
android:textOn	用于设置开关按钮被选中时的说明文本

继承于 android.widget.TextView 类的 ToggleButton 的常用属性如表 2.28 所示。

表 2.28 继承于 android.widget.TextView 类的 ToggleButton 的常用属性

属性	描述
android:autoText	若被选中，则表示对开关按钮的说明文本自动更正拼写错误
android:drawableBottom	用于在开关按钮的说明文本的底端绘制指定图片，该图片可以是放在 res/drawable 目录下的图片，通过"@drawable/文件名（不包括文件的扩展名）"设置
android:drawableRight	用于在开关按钮的说明文本的右侧绘制指定图片，该图片可以是放在 res/drawable 目录下的图片，通过"@drawable/文件名（不包括文件的扩展名）"设置
android:editable	用于设置开关按钮的说明文本能否被编辑，默认值为 true
android:text	用于设置开关按钮的默认说明文本

继承于 android.view.View 类的 ToggleButton 的常用属性如表 2.29 所示。

表 2.29 继承于 android.view.View 类的 ToggleButton 的常用属性

属性	描述
android:background	用于设置开关按钮的背景图片，该图片可以是放在 res/drawable 目录下的图片，通过"@drawable/文件名（不包括文件的扩展名）"设置
android:contentDescription	用于设置开关按钮的简单描述文字
android:id	用于设置开关按钮的名称
android:onClick	用于设置单击事件响应方法的方法名
android:visibility	用于设置开关按钮是否可见

【例 2.15】 ToggleButton 实例

这个例子将使用 ToggleButton 来创建一个简单的 Android 应用,下面将通过表 2.30 中的步骤来实现。

表 2.30 ToggleButton 实例步骤

步 骤	描 述
1	打开 Android Studio,创建一个 Android 应用,选择"Empty Activity"选项,在"Name"文本框中输入项目名,在"Package name"文本框中输入包名
2	在"Minimum SDK"下拉列表中选择"API 18: Android 4.3(Jelly Bean)"选项或其他 SDK
3	在"Language"下拉列表中选择"Java"选项
4	选择保存路径,单击"Finish"按钮
5	在工程项目中找到 ras/layout 目录中的 activity_main.xml 文件,并在其中添加一个 TextView、两个 CheckBox、两个 Button 和一个 ImageButton 控件,按如图 2.24 所示的运行效果进行摆放
6	将工程项目的 res/mipmap 目录下的图片文件 ic_launcher.png 复制、粘贴到 res/drawable 目录下
7	启动 Genymotion 模拟器,然后在 Android 工程项目中进行如下代码修改

修改新建工程项目的 res/layout 目录下的布局文件 activity_main.xml,修改代码如下:

```xml
<?xml version="1.0" encoding="utf-8"?>
<RelativeLayout xmlns:android="http://schemas.android.com/apk/res/android"
    xmlns:tools="http://schemas.android.com/tools"
    android:layout_width="match_parent"
    android:layout_height="match_parent"
    android:paddingLeft="@dimen/activity_horizontal_margin"
    android:paddingRight="@dimen/activity_horizontal_margin"
    android:paddingTop="@dimen/activity_vertical_margin"
    android:paddingBottom="@dimen/activity_vertical_margin"
    tools:context=".MainActivity">

    <TextView
        android:id="@+id/textView1"
        android:layout_width="wrap_content"
        android:layout_height="wrap_content"
        android:text="Tutorials point"
        android:textColor="#ff87ff09"
        android:textSize="30dp"
        android:layout_above="@+id/imageButton"
        android:layout_centerHorizontal="true"
        android:layout_marginBottom="40dp" />

    <ImageButton
        android:layout_width="wrap_content"
        android:layout_height="wrap_content"
        android:id="@+id/imageButton"
        android:src="@drawable/ic_launcher"
        android:layout_centerVertical="true"
        android:layout_centerHorizontal="true" />
```

```xml
<ToggleButton
    android:layout_width="wrap_content"
    android:layout_height="wrap_content"
    android:text="ON"
    android:id="@+id/toggleButton"
    android:checked="true"
    android:layout_below="@+id/imageButton"
    android:layout_toLeftOf="@+id/imageButton"
    android:layout_toStartOf="@+id/imageButton" />

<ToggleButton
    android:layout_width="wrap_content"
    android:layout_height="wrap_content"
    android:text="OFF"
    android:id="@+id/toggleButton2"
    android:checked="true"
    android:layout_below="@+id/imageButton"
    android:layout_toEndOf="@+id/imageButton" />

<Button
    android:layout_width="wrap_content"
    android:layout_height="wrap_content"
    android:id="@+id/button2"
    android:text="CLICKME"
    android:layout_alignParentBottom="true"
    android:layout_centerHorizontal="true" />
</RelativeLayout>
```

修改新建工程项目的 java/com.example.togglebutton 目录下的 Java 文件 MainActivity.java，修改代码如下：

```java
package com.example.togglebutton;

import android.os.Bundle;
import android.support.v7.app.AppCompatActivity;
import android.view.View;
import android.widget.Button;
import android.widget.Toast;
import android.widget.ToggleButton;

public class MainActivity extends AppCompatActivity {
    ToggleButton tg1,tg2;
    Button b1;
    @Override
    protected void onCreate(Bundle savedInstanceState) {
        super.onCreate(savedInstanceState);
        setContentView(R.layout.activity_main);
        tg1=(ToggleButton)findViewById(R.id.toggleButton);
        tg2=(ToggleButton)findViewById(R.id.toggleButton2);
```

```
            b1=(Button)findViewById(R.id.button2);
            b1.setOnClickListener(new View.OnClickListener() {
                @Override
                public void onClick(View v) {
                    StringBuffer result = new StringBuffer();
                    result.append("You have clicked first ON Button-:) ").
                            append(tg1.getText());
                    result.append("You have clicked Second ON Button -:) ").
                            append(tg2.getText());
                    Toast.makeText(MainActivity.this,result.toString(),Toast.
                            LENGTH_SHORT).show();
                }
            });
        }
    }
```

在 Genymotion 中运行程序，然后单击其中一个 ToggleButton 控件，使其状态变为 OFF，并单击"CLICKME"按钮，效果如图 2.24 所示。

图 2.24　ToggleButton 实例运行效果

2.4.8　RadioButton（单选按钮）与 RadioGroup（按钮组）

在默认情况下，单选按钮显示为一个圆形图标，并且在该图标旁边会放置一些说明文字。一般将多个单选按钮放置在按钮组中，当用户选中其中某个单选按钮后，按钮组中的其他按钮将被自动取消选中状态。RadioButton 是 Button 类的子类，所以它可以直接使用 Button 的各种属性。

下面介绍 RadioButton 的常用属性，关于该控件的其他属性，可以参阅 Android 官方提供的完整的 API 文档。

（1）继承于 android.widget.TextView 类的 RadioButton 的常用属性如表 2.31 所示。

表 2.31　继承于 android.widget.TextView 类的 RadioButton 的常用属性

属　　性	描　　述
android:autoText	若被选中，则表示对单选按钮的说明文本自动更正拼写错误
android:drawableBottom	用于在单选按钮的说明文本的底端绘制指定图片，该图片可以是放在 res/drawable 目录下的图片，通过"@drawable/文件名（不包括文件的扩展名）"设置
android:drawableRight	用于在单选按钮的说明文本的右侧绘制指定图片，该图片可以是放在 res/drawable 目录下的图片，通过"@drawable/文件名（不包括文件的扩展名）"设置
android:enabled	用于设置单选按钮是否可用
android:text	用于设置单选按钮的说明文本

（2）继承于 android.view.View 类的 RadioButton 的常用属性如表 2.32 所示。

表 2.32　继承于 android.view.View 类的 RadioButton 的常用属性

属　　性	描　　述
android:background	用于设置单选按钮的背景图片，该图片可以是放在 res/drawable 目录下的图片，通过"@drawable/文件名（不包括文件的扩展名）"设置
android:contentDescription	用于设置单选按钮的简单描述文字
android:id	用于设置单选按钮的名称
android:onClick	用于设置单击事件响应方法的方法名
android:visibility	用于设置单选按钮是否可见

【例 2.16】　RadioButton 实例

这个例子将使用 RadioButton 来创建一个简单的 Android 应用，下面将通过表 2.33 中的步骤来实现。

表 2.33　RadioButton 实例步骤

步　　骤	描　　述
1	打开 Android Studio，创建一个 Android 应用，选择"Empty Activity"选项，在"Name"文本框中输入项目名，在"Package name"文本框中输入包名
2	在"Minimum SDK"下拉列表中选择"API 18: Android 4.3（Jelly Bean）"选项或其他 SDK
3	在"Language"下拉列表中选择"Java"选项
4	选择保存路径，单击"Finish"按钮
5	在工程项目中找到 ras/layout 目录中的 activity_main.xml 文件，并在其中添加一个 TextView、一个 ImageButton、一个 Button 和一个 RadioGroup 控件，在 RadioGroup 中放置 3 个 RadioButton，按如图 2.25 所示的运行效果进行摆放
6	将工程项目的 res/mipmap 目录下的图片文件 ic_launcher.png 复制、粘贴到 res/drawable 目录下
7	启动 Genymotion 模拟器，然后在 Android 工程项目中进行如下代码修改

修改新建工程项目的 res/layout 目录下的布局文件 activity_main.xml，修改代码如下：

```
<?xml version="1.0" encoding="utf-8"?>
<RelativeLayout xmlns:android="http://schemas.android.com/apk/res/android"
    xmlns:tools="http://schemas.android.com/tools"
    android:layout_width="match_parent"
    android:layout_height="match_parent"
    android:paddingLeft="@dimen/activity_horizontal_margin"
    android:paddingRight="@dimen/activity_horizontal_margin"
    android:paddingTop="@dimen/activity_vertical_margin"
```

```xml
        android:paddingBottom="@dimen/activity_vertical_margin"
        tools:context=".MainActivity">

        <TextView
            android:id="@+id/textView1"
            android:layout_width="wrap_content"
            android:layout_height="wrap_content"
            android:text="Example of Radio Button"
            android:layout_alignParentTop="true"
            android:layout_centerHorizontal="true"
            android:textSize="30dp" />

        <ImageButton
            android:layout_width="wrap_content"
            android:layout_height="wrap_content"
            android:id="@+id/imageButton"
            android:src="@drawable/ic_launcher"
            android:layout_below="@+id/textView1"
            android:layout_centerHorizontal="true" />

        <Button
            android:layout_width="wrap_content"
            android:layout_height="wrap_content"
            android:id="@+id/button2"
            android:text="CLICKME"
            android:layout_alignParentBottom="true"
            android:layout_centerHorizontal="true" />

        <RadioGroup

            android:layout_width="fill_parent"
            android:layout_height="fill_parent"
            android:layout_below="@+id/imageButton"
            android:id="@+id/radiogroup1"
            android:layout_centerHorizontal="true">

            <RadioButton
                android:layout_width="142dp"
                android:layout_height="wrap_content"
                android:text="JAVA"
                android:id="@+id/radioButton"
                android:textSize="25dp"
                android:textColor="@android:color/holo_red_light"
                android:checked="false"
                android:layout_gravity="center_horizontal" />

            <RadioButton
                android:layout_width="wrap_content"
                android:layout_height="wrap_content"
```

```
            android:text="ANDROID"
            android:id="@+id/radioButton2"
            android:layout_gravity="center_horizontal"
            android:checked="false"
            android:textColor="@android:color/holo_red_dark"
            android:textSize="25dp" />

        <RadioButton
            android:layout_width="136dp"
            android:layout_height="wrap_content"
            android:text="HTML"
            android:id="@+id/radioButton3"
            android:layout_gravity="center_horizontal"
            android:checked="false"
            android:textSize="25dp"
            android:textColor="@android:color/holo_red_dark" />

    </RadioGroup>
</RelativeLayout>
```

修改新建工程项目的 java/com.example.radiobutton 目录下的 Java 文件 MainActivity.java，修改代码如下：

```
package com.example.radiobutton;

import android.os.Bundle;
import android.support.v7.app.AppCompatActivity;
import android.view.View;
import android.widget.Button;
import android.widget.RadioButton;
import android.widget.RadioGroup;
import android.widget.Toast;

public class MainActivity extends AppCompatActivity {
    RadioGroup rg1;
    RadioButton rb1;
    Button b1;
    @Override
    protected void onCreate(Bundle savedInstanceState) {
        super.onCreate(savedInstanceState);
        setContentView(R.layout.activity_main);
        addListenerRadioButton();
    }
    private void addListenerRadioButton() {
        rg1 = (RadioGroup) findViewById(R.id.radiogroup1);
        b1 = (Button) findViewById(R.id.button2);
        b1.setOnClickListener(new View.OnClickListener() {
            @Override
            public void onClick(View v) {
```

```
                    int selected = rg1.getCheckedRadioButtonId();
                    rb1 = (RadioButton) findViewById(selected);
                    Toast.makeText(MainActivity.this, rb1.getText(), Toast.
                            LENGTH_LONG).show();
                }
            });
        }
    }
```

在 Genymotion 中运行程序,然后选中其中一个 RadioButton 控件,并单击 "CLICKME" 按钮,效果如图 2.25 所示。

图 2.25　RadioButton 与 RadioGroup 实例运行效果

2.4.9　使用 ProgressDialog（进度对话框）类创建 ProgressBar（进度条）

当一个应用程序在后台执行时,前台界面不会有任何信息,这时用户不知道程序是否在执行及执行进度等信息,因此需要使用进度条来提示程序执行的进度。例如,当用户从互联网上传或下载文件时,上传/下载进度条将显示下载进度信息。

在 Android 中,可以使用 ProgressDialog 类来创建 ProgressBar。首先使用 ProgressDialog 类实例化一个对象,语法如下:

```
ProgressDialog progress = new ProgressDialog(this);
```

然后设置 ProgressDialog 类的属性,代码如下:

```
progress.setMessage("Downloading Music ! ");
progress.setProgressStyle(ProgressDialog.STYLE_HORIZONTAL);
progress.setIndeterminate(true);
```

ProgressDialog 类的常用方法如表 2.34 所示。

表 2.34 ProgressDialog 类的常用方法

方法	描述
getMax()	此方法将返回进度条的最大值
incrementProgressBy(int diff)	此方法将设置增加的进度，每次推进的步伐
setIndeterminate(boolean indeterminate)	此方法将设置进度条是否为不确定模式
setMax(int max)	此方法将设置进度条的最大值
setProgress(int value)	此方法将设置进度条的当前进度值
show(Context context, CharSequence title, CharSequence message)	这是一个静态方法，用于显示进度对话框

【例 2.17】 ProgressBar 实例

这个例子将使用 ProgressDialog 和 ProgressBar 来创建一个简单的 Android 应用，下面将通过表 2.35 中的步骤来实现。

表 2.35 ProgressDialog/ProgressBar 实例步骤

步骤	描述
1	打开 Android Studio，创建一个 Android 应用，选择"Empty Activity"选项，在"Name"文本框中输入项目名，在"Package name"文本框中输入包名
2	在"Minimum SDK"下拉列表中选择"API 18: Android 4.3（Jelly Bean）"选项或其他 SDK
3	在"Language"下拉列表中选择"Java"选项
4	选择保存路径，单击"Finish"按钮
5	在工程项目中找到 ras/layout 目录中的 activity_main.xml 文件，并在其中添加一个 TextView 和一个 Button 控件，按如图 2.26 所示的运行效果进行摆放
6	启动 Genymotion 模拟器，然后在 Android 工程项目中进行如下代码修改

修改新建工程项目的 res/layout 目录下的布局文件 activity_main.xml，修改代码如下：

```xml
<?xml version="1.0" encoding="utf-8"?>
<RelativeLayout xmlns:android="http://schemas.android.com/apk/res/android"
    xmlns:tools="http://schemas.android.com/tools"
    android:layout_width="match_parent"
    android:layout_height="match_parent"
    android:paddingLeft="@dimen/activity_horizontal_margin"
    android:paddingRight="@dimen/activity_horizontal_margin"
    android:paddingTop="@dimen/activity_vertical_margin"
    android:paddingBottom="@dimen/activity_vertical_margin"
    tools:context=".MainActivity">


    <TextView
        android:layout_width="wrap_content"
        android:layout_height="wrap_content"
        android:id="@+id/textView"
        android:layout_alignParentTop="true"
        android:layout_centerHorizontal="true"
        android:textSize="30dp"
        android:text="Progress bar" />

    <Button
        android:layout_width="wrap_content"
```

```
                android:layout_height="wrap_content"
                android:text="DOWNLOAD"
                android:onClick="download"
                android:id="@+id/button2"
                android:layout_marginLeft="125dp"
                android:layout_marginStart="125dp"
                android:layout_centerVertical="true" />
</RelativeLayout>
```

修改新建工程项目的 java/com.example.progressbar 目录下的 Java 文件 MainActivity.java，修改代码如下：

```
package com.example.progressbar;

import android.app.ProgressDialog;
import android.os.Bundle;
import android.support.v7.app.AppCompatActivity;
import android.view.View;
import android.widget.Button;

public class MainActivity extends AppCompatActivity {
    Button b1;
    private ProgressDialog progress;
    @Override
    protected void onCreate(Bundle savedInstanceState) {
        super.onCreate(savedInstanceState);
        setContentView(R.layout.activity_main);
        b1 = (Button) findViewById(R.id.button2);
    }
    public void download(View view){
        progress=new ProgressDialog(this);
        progress.setMessage("Downloading Music");
        progress.setProgressStyle(ProgressDialog.STYLE_HORIZONTAL);
        progress.setIndeterminate(true);
        progress.setProgress(0);
        progress.show();

        final int totalProgressTime = 100;
        final Thread t = new Thread() {
            @Override
            public void run() {
                int jumpTime = 0;

                while(jumpTime < totalProgressTime) {
                    try {
                        sleep(200);
                        jumpTime += 5;
                        progress.setProgress(jumpTime);
                    }
                    catch (InterruptedException e) {
```

```
                            // TODO Auto-generated catch block
                            e.printStackTrace();
                        }
                    }
                }
            };
            t.start();
        }
    }
```

程序在 Genymotion 中的运行效果如图 2.26 所示。

当用户单击"DOWNLOAD"按钮后将弹出进度条对话框,程序的运行效果如图 2.27 所示。

图 2.26 ProgressDialog/ProgressBar
　　　　实例运行效果（1）

图 2.27 ProgressDialog/ProgressBar
　　　　实例运行效果（2）

2.4.10 Spinner（列表选择框）

图 2.28 Spinner 实例

Android 中提供的 Spinner 相当于在网页中常见的下拉列表框,用于供用户选择相应选项,从而方便用户使用。在用户使用电子邮件时,可以选择 Reply（回复）、Reply all（回复全部）和 Forward（转发）菜单项,如图 2.28 所示。

【例 2.18】 Spinner 实例

这个例子将使用 Spinner 来创建一个简单的 Android 应用,下面将通过表 2.36 中的步骤来实现。

表 2.36 Spinner 实例步骤

步　　骤	描　　述
1	打开 Android Studio,创建一个 Android 应用,选择"Empty Activity"选项,在"Name"文本框中输入项目名,在"Package name"文本框中输入包名

步骤	描述
2	在"Minimum SDK"下拉列表中选择"API 18: Android 4.3（Jelly Bean）"选项或其他 SDK
3	在"Language"下拉列表中选择"Java"选项
4	选择保存路径，单击"Finish"按钮
5	在工程项目中找到 ras/layout 目录中的 activity_main.xml 文件，并在其中添加一个 TextView 和一个 Spinner 控件，按如图 2.29 所示的运行效果进行摆放
6	启动 Genymotion 模拟器，然后在 Android 工程项目中进行如下代码修改

修改新建工程项目的 res/layout 目录下的布局文件 activity_main.xml，修改代码如下：

```xml
<?xml version="1.0" encoding="utf-8"?>
<LinearLayout xmlns:android="http://schemas.android.com/apk/res/android"
    android:orientation="vertical"
    android:padding="10dip"
    android:layout_width="fill_parent"
    android:layout_height="wrap_content">

    <TextView
        android:layout_width="fill_parent"
        android:layout_height="wrap_content"
        android:layout_marginTop="10dip"
        android:text="Category:"
        android:layout_marginBottom="5dp" />

    <Spinner
        android:id="@+id/spinner"
        android:layout_width="fill_parent"
        android:layout_height="wrap_content" />

</LinearLayout>
```

修改新建工程项目的 java/com.example.spinner 目录下的 Java 文件 MainActivity.java，修改代码如下：

```java
package com.example.spinner;

import android.app.Activity;
import android.os.Bundle;
import android.view.View;
import android.widget.AdapterView;
import android.widget.AdapterView.OnItemSelectedListener;
import android.widget.ArrayAdapter;
import android.widget.Spinner;
import android.widget.Toast;

import java.util.ArrayList;
import java.util.List;

public class MainActivity extends Activity implements OnItemSelectedListener {
```

```java
@Override
protected void onCreate(Bundle savedInstanceState) {
    super.onCreate(savedInstanceState);
    setContentView(R.layout.activity_main);
    // Spinner element
    Spinner spinner = (Spinner) findViewById(R.id.spinner);

    // Spinner click listener
    spinner.setOnItemSelectedListener(this);

    // Spinner Drop down elements
    List<String> categories = new ArrayList<String>();
    categories.add("Automobile");
    categories.add("Business Services");
    categories.add("Computers");
    categories.add("Education");
    categories.add("Personal");
    categories.add("Travel");

    // Creating adapter for spinner
    ArrayAdapter<String> dataAdapter = new ArrayAdapter<String>(this,
            android.R.layout.simple_spinner_item, categories);

    // Drop down layout style - list view with radio button
    dataAdapter.setDropDownViewResource(android.R.layout.simple_
            spinner_dropdown_item);

    // attaching data adapter to spinner
    spinner.setAdapter(dataAdapter);
}

@Override
public void onItemSelected(AdapterView<?> parent, View view, int
    position, long id) {
    // On selecting a spinner item
    String item = parent.getItemAtPosition(position).toString();

    // Showing selected spinner item
    Toast.makeText(parent.getContext(), "Selected: " + item, Toast.
            LENGTH_LONG).show();
}
public void onNothingSelected(AdapterView<?> arg0) {
    // TODO Auto-generated method stub
}

}
```

程序在 Genymotion 中的运行效果如图 2.29 所示。

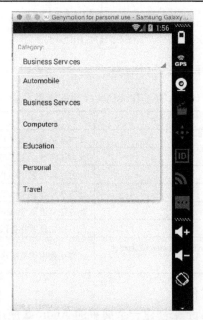

图 2.29　Spinner 实例运行效果

2.4.11　TimePicker（时间拾取器）

在 Android 中，TimePicker 是用于选择一天中时间的控件，在 TimePicker 中可以设置小时、分钟和 AM/PM（上午/下午），如图 2.30 所示。

图 2.30　TimePicker 实例

为了使用 TimePicker 类，首先需要在布局文件 activity.xml 中定义 TimePicker，语法如下：

```
<TimePicker
    android:id="@+id/timePicker1"
    android:layout_width="wrap_content"
    android:layout_height="wrap_content" />
```

在定义 TimePicker 后，需要在 Java 文件中实例化这个控件，实例化 TimePicker 的语法如下：

```
import android.widget.TimePicker;
private TimePicker timePicker1;
timePicker1 = (TimePicker) findViewById(R.id.timePicker1);
```

在实例化控件后，可以通过 getCurrentHour()方法和 getCurrentMinute()方法获取当前系统的小时和分钟数，语法如下：

```
int hour = timePicker1.getCurrentHour();
int min = timePicker1.getCurrentMinute();
```

TimePicker 类的常用方法如表 2.37 所示。

表 2.37　TimePicker 类的常用方法

方　　法	描　　述
is24HourView()	此方法将返回当前 TimePicker 是否以 24 小时模式显示的状态
isEnabled()	此方法将返回当前 TimePicker 是否可用的状态
setCurrentHour(Integer currentHour)	此方法将设置 TimePicker 中显示的当前小时数
setCurrentMinute(Integer currentMinute)	此方法将设置 TimePicker 中显示的当前分钟数
setEnabled(boolean enabled)	此方法将设置 TimePicker 是否可用
setIs24HourView(Boolean is24HourView)	此方法将设置是否以 24 小时模式显示时间
setOnTimeChangedListener(TimePicker.OnTimeChangedListener onTimeChangedListener)	此方法将用于监听事件改变时触发的事件

【例 2.19】　TimePicker 实例

这个例子将使用 TimePicker 来创建一个简单的 Android 应用，下面将通过表 2.38 中的步骤来实现。

表 2.38　TimePicker 实例步骤

步　骤	描　　述
1	打开 Android Studio，创建一个 Android 应用，选择 "Empty Activity" 选项，在 "Name" 文本框中输入项目名，在 "Package name" 文本框中输入包名
2	在 "Minimum SDK" 下拉列表中选择 "API 18: Android 4.3（Jelly Bean）" 选项或其他 SDK
3	在 "Language" 下拉列表中选择 "Java" 选项
4	选择保存路径，单击 "Finish" 按钮
5	在工程项目中找到 ras/layout 目录中的 activity_main.xml 文件，并在其中添加 3 个 TextView、一个 Button 和一个 TimePicker 控件，按如图 2.31 所示的运行效果进行摆放
6	启动 Genymotion 模拟器，然后在 Android 工程项目中进行如下代码修改

修改新建工程项目的 res/layout 目录下的布局文件 activity_main.xml，修改代码如下：

```
<?xml version="1.0" encoding="utf-8"?>
<RelativeLayout xmlns:android="http://schemas.android.com/apk/res/android"
    xmlns:tools="http://schemas.android.com/tools"
    android:layout_width="match_parent"
    android:layout_height="match_parent"
    android:paddingLeft="@dimen/activity_horizontal_margin"
    android:paddingRight="@dimen/activity_horizontal_margin"
    android:paddingTop="@dimen/activity_vertical_margin"
    android:paddingBottom="@dimen/activity_vertical_margin"
    tools:context=".MainActivity">

    <TextView
        android:id="@+id/textView2"
        android:layout_width="wrap_content"
        android:layout_height="wrap_content"
```

```xml
            android:layout_alignParentTop="true"
            android:layout_centerHorizontal="true"
            android:text="@string/time_pick"
            android:textAppearance="?android:attr/textAppearanceMedium" />

        <Button
            android:id="@+id/set_button"
            android:layout_width="wrap_content"
            android:layout_height="wrap_content"
            android:layout_alignParentBottom="true"
            android:layout_centerHorizontal="true"
            android:layout_marginBottom="180dp"
            android:onClick="setTime"
            android:text="@string/time_save" />

        <TimePicker
            android:id="@+id/timePicker1"
            android:layout_width="wrap_content"
            android:layout_height="wrap_content"
            android:layout_above="@+id/set_button"
            android:layout_centerHorizontal="true"
            android:layout_marginBottom="24dp" />

        <TextView
            android:id="@+id/textView3"
            android:layout_width="wrap_content"
            android:layout_height="wrap_content"
            android:layout_alignLeft="@+id/timePicker1"
            android:layout_alignTop="@+id/set_button"
            android:layout_marginTop="67dp"
            android:text="@string/time_current"
            android:textAppearance="?android:attr/textAppearanceMedium" />

        <TextView
            android:id="@+id/textView1"
            android:layout_width="wrap_content"
            android:layout_height="wrap_content"
            android:layout_below="@+id/textView3"
            android:layout_centerHorizontal="true"
            android:text="@string/time_selected"
            android:textAppearance="?android:attr/textAppearanceMedium" />
    </RelativeLayout>
```

修改新建工程项目的 res/values/ 目录下的数组文件 strings.xml，修改代码如下：

```xml
<resources>
    <string name="app_name">TimePicker</string>
    <string name="time_picker_example">Time Picker Example</string>
    <string name="time_pick">Pick the time and press save button</string>
    <string name="time_save">SAVE</string>
```

```xml
        <string name="time_selected"></string>
        <string name="time_current">The Time is:</string>
</resources>
```

修改新建工程项目的 java/com.example.timepicker 目录下的 Java 文件 MainActivity.java，修改代码如下：

```java
package com.example.timepicker;

import android.os.Bundle;
import android.support.v7.app.AppCompatActivity;
import android.view.View;
import android.widget.TextView;
import android.widget.TimePicker;

import java.util.Calendar;

public class MainActivity extends AppCompatActivity {
    private TimePicker timePicker1;
    private TextView time;
    private Calendar calendar;
    private String format = "";
    @Override
    protected void onCreate(Bundle savedInstanceState) {
        super.onCreate(savedInstanceState);
        setContentView(R.layout.activity_main);
        timePicker1 = (TimePicker) findViewById(R.id.timePicker1);
        time = (TextView) findViewById(R.id.textView1);
        calendar = Calendar.getInstance();

        int hour = calendar.get(Calendar.HOUR_OF_DAY);
        int min = calendar.get(Calendar.MINUTE);
        showTime(hour, min);
    }

    public void setTime(View view) {
        int hour = timePicker1.getCurrentHour();
        int min = timePicker1.getCurrentMinute();
        showTime(hour, min);
    }

    public void showTime(int hour, int min) {
        if (hour == 0) {
            hour += 12;
            format = "AM";
        }
        else if (hour == 12) {
            format = "PM";
        } else if (hour > 12) {
            hour -= 12;
```

```
                format = "PM";
            } else {
                format = "AM";
            }
            time.setText(new StringBuilder().append(hour).append(":").append(min)
                    .append(" ").append(format));
        }
    }
```

程序在 Genymotion 中的运行效果如图 2.31 所示。

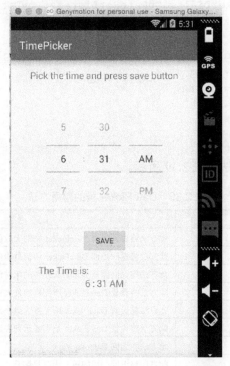

图 2.31 TimePicker 实例运行效果

2.4.12 DatePicker（日期拾取器）与 DatePickerDialog（日期拾取器对话框）

在 Android 中，DatePicker 是让用户在界面中选择日期的控件。DatePicker 由年、月、日 3 个部分组成。Android 提供了使用 DatePicker 和调用 DatePickerDialog 类两种方式来实现日期选取功能。

在本节中，将通过调用 DatePickerDialog 类的方式来实现日期选取功能。若要在屏幕上显示 DatePickerDialog，则需要通过 showDialog(id_of_dialog)来调用，其语法如下：

```
showDialog(999);    // 其中 999 为 DatePickerDialog id
```

在调用 showDialog()方法后，onCreateDialog()方法将自动被调用，所以需要重写 onCreateDialog()方法，其语法如下：

```
@Override
```

```
protected Dialog onCreateDialog(int id) {
    // TODO Auto-generated method stub
    if (id == 999) {
        return new DatePickerDialog(this, myDateListener, year, month, day);
    }
    return null;
}
```

最后一步，需要注册 DatePickerDialog 监听器并重写 onDateSet()方法。onDateSet()方法包含更新后的年、月、日数据。DatePickerDialog 监听器与 onDateSet()方法的语法如下：

```
private DatePickerDialog.OnDateSetListener myDateListener = new
        DatePickerDialog.OnDateSetListener() {
    @Override
    public void onDateSet(DatePicker arg0, int arg1, int arg2, int arg3) {
        // arg1 = year
        // arg2 = month
        // arg3 = day
    }
};
```

DatePicker 类的常用方法如表 2.39 所示。

表 2.39　DatePicker 类的常用方法

方法	描述
getDayOfMonth()	此方法将返回当前 DatePicker 中的天数值
getMonth()	此方法将返回当前 DatePicker 中的月份值
getYear()	此方法将返回当前 DatePicker 中的年份值
setMaxDate(long maxDate)	此方法将设置 DatePicker 所支持的最大日期数
setMinDate(long minDate)	此方法将设置 DatePicker 所支持的最小日期数
setSpinnersShown(boolean shown)	此方法将设置 DatePicker 是否显示列表选择框
getCalendarView()	此方法将返回 CalendarView
getFirstDayOfWeek()	此方法将返回一周的第一天是星期几
updateDate(int year, int month, int dayOfMonth)	此方法将通过参数更新 DatePicker 的现有时间

【例 2.20】 DatePicker 实例

这个例子将使用 DatePicker 来创建一个简单的 Android 应用，下面将通过表 2.40 中的步骤来实现。

表 2.40　DatePicker 实例步骤

步骤	描述
1	打开 Android Studio，创建一个 Android 应用，选择"Empty Activity"选项，在"Name"文本框中输入项目名，在"Package name"文本框中输入包名
2	在"Minimum SDK"下拉列表中选择"API 18: Android 4.3（Jelly Bean）"选项或其他 SDK
3	在"Language"下拉列表中选择"Java"选项
4	选择保存路径，单击"Finish"按钮

步骤	描述
5	在工程项目中找到 ras/layout 目录中的 activity_main.xml 文件,并在其中添加 3 个 TextView 和一个 Button 控件,按如图 2.32 所示的运行效果进行摆放
6	启动 Genymotion 模拟器,然后在 Android 工程项目中进行如下代码修改

修改新建工程项目的 res/layout 目录下的布局文件 activity_main.xml,修改代码如下:

```xml
<?xml version="1.0" encoding="utf-8"?>
<RelativeLayout xmlns:android="http://schemas.android.com/apk/res/android"
    xmlns:tools="http://schemas.android.com/tools"
    android:layout_width="match_parent"
    android:layout_height="match_parent"
    android:paddingLeft="@dimen/activity_horizontal_margin"
    android:paddingRight="@dimen/activity_horizontal_margin"
    android:paddingTop="@dimen/activity_vertical_margin"
    android:paddingBottom="@dimen/activity_vertical_margin"
    tools:context=".MainActivity">

    <Button
        android:id="@+id/button1"
        android:layout_width="wrap_content"
        android:layout_height="wrap_content"
        android:layout_alignParentTop="true"
        android:layout_centerHorizontal="true"
        android:layout_marginTop="70dp"
        android:onClick="setDate"
        android:text="@string/date_button_set" />

    <TextView
        android:id="@+id/textView1"
        android:layout_width="wrap_content"
        android:layout_height="wrap_content"
        android:layout_alignParentTop="true"
        android:layout_centerHorizontal="true"
        android:layout_marginTop="24dp"
        android:text="@string/date_label_set"
        android:textAppearance="?android:attr/textAppearanceMedium" />

    <TextView
        android:id="@+id/textView2"
        android:layout_width="wrap_content"
        andraoid:layout_height="wrap_content"
        android:layout_below="@+id/button1"
        android:layout_marginTop="66dp"
        android:layout_toLeftOf="@+id/button1"
        android:text="@string/date_view_set"
        android:textAppearance="?android:attr/textAppearanceMedium" />
```

```xml
<TextView
    android:id="@+id/textView3"
    android:layout_width="wrap_content"
    android:layout_height="wrap_content"
    android:layout_alignRight="@+id/button1"
    android:layout_below="@+id/textView2"
    android:layout_marginTop="72dp"
    android:text="@string/date_selected"
    android:textAppearance="?android:attr/textAppearanceMedium" />
</RelativeLayout>
```

修改新建工程项目的 res/values/目录下的数组文件 strings.xml，修改代码如下：

```xml
<resources>
    <string name="app_name">DatePicker</string>
    <string name="hello_world">Hello World!</string>
    <string name="date_label_set">Press the button to set the date</string>
    <string name="date_button_set">SET DATE</string>
    <string name="date_view_set">The Date is: </string>
    <string name="date_selected"></string>
</resources>
```

修改新建工程项目的 java/com.example.datapicker 目录下的 Java 文件 MainActivity.java，修改代码如下：

```java
package com.example.datepicker;

import android.app.DatePickerDialog;
import android.app.Dialog;
import android.os.Bundle;
import android.support.v7.app.AppCompatActivity;
import android.view.View;
import android.widget.DatePicker;
import android.widget.TextView;
import android.widget.Toast;

import java.util.Calendar;

public class MainActivity extends AppCompatActivity {
    private DatePicker datePicker;
    private Calendar calendar;
    private TextView dateView;
    private int year, month, day;
    @Override
    protected void onCreate(Bundle savedInstanceState) {
        super.onCreate(savedInstanceState);
        setContentView(R.layout.activity_main);
```

```java
        dateView = (TextView) findViewById(R.id.textView3);
        calendar = Calendar.getInstance();
        year = calendar.get(Calendar.YEAR);

        month = calendar.get(Calendar.MONTH);
        day = calendar.get(Calendar.DAY_OF_MONTH);
        showDate(year, month + 1, day);
    }
    @SuppressWarnings("deprecation")
    public void setDate(View view) {
        showDialog(999);
        Toast.makeText(getApplicationContext(), "ca", Toast.LENGTH_SHORT)
                .show();
    }

    @Override
    protected Dialog onCreateDialog(int id) {
        // TODO Auto-generated method stub
        if (id == 999) {
            return new DatePickerDialog(this, myDateListener, year, month, day);
        }
        return null;
    }

    private DatePickerDialog.OnDateSetListener myDateListener = new
            DatePickerDialog.OnDateSetListener() {
        @Override
        public void onDateSet(DatePicker arg0, int arg1, int arg2, int arg3) {
            // TODO Auto-generated method stub
            // arg1 = year
            // arg2 = month
            // arg3 = day
            showDate(arg1, arg2+1, arg3);
        }
    };

    private void showDate(int year, int month, int day) {
        dateView.setText(new StringBuilder().append(day).append("/")
                .append(month).append("/").append(year));
    }
}
```

程序在 Genymotion 中的运行效果如图 2.32 所示。

当用户单击"SET DATE"按钮后,将弹出日期拾取器对话框,程序的运行效果如图 2.33 所示。

图 2.32 DatePicker 实例运行效果（1）　　　图 2.33 DatePicker 实例运行效果（2）

第3章 基本程序单元 Activity

在 Android 中，Activity（活动）表示一个单一屏幕上的用户界面。例如，电子邮件应用程序的主界面可能是一个 Activity，显示新的电子邮件列表的界面则是另一个 Activity，撰写电子邮件的界面又是其他的 Activity。如果应用程序有一个以上的 Activity，则应该在 AndroidManifest.xml 配置文件中将其中一个 Activity 标记为启动 Activity 以开启应用程序。

如果用户曾经使用过 C、C++和 Java，就会知道程序一般是从 main()函数开始运行的。相似地，Android 系统是从 MainActivity 的 onCreate()方法调用开始启动程序的。在图 3.1 中，矩形方框表示可以被回调的方法，椭圆形表示 Activity 的状态。从该图中可以看出，在一个 Activity 的生命周期中有一些方法会被系统调用。

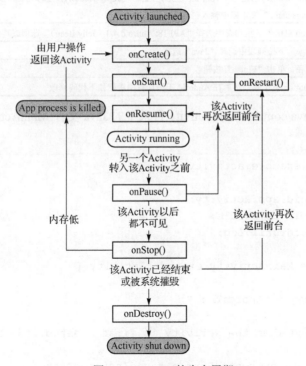

图 3.1 Activity 的生命周期

Activity 生命周期中的回调方法如表 3.1 所示。

表 3.1 Activity 生命周期中的回调方法

回调方法	描述
onCreate()	该方法在创建 Activity 时被调用，是非常常见的方法。在 Android Studio 中创建项目时，会自动创建一个 MainActivity，在该 MainActivity 中，默认重写了 onCreate（Bundle savedInstanceState）方法，用于对该 Activity 进行初始化
onStart()	该方法在启动 Activity 时被回调，也就是当一个 Activity 变为显示时被回调
onResume()	该方法在将 Activity 由暂停状态恢复为活动状态时被调用。调用该方法后，该 Activity 位于 Activity 堆栈的栈顶。该方法总是在 onPause()方法后被执行

回调方法	描述
onPause()	该方法在暂停 Activity 时被调用。该方法需要被非常快地执行,并且直到该方法被执行完毕后,下一个 Activity 才能被恢复。该方法通常用于保存数据。比如,当用户玩游戏时,突然来了一个电话,这时使用该方法保存游戏状态的数据
onStop()	该方法在停止 Activity 时被回调
onDestroy()	该方法在销毁 Activity 时被回调
onRestart()	该方法在重新启动 Activity 时被回调,并且总是在 onStart()方法后被执行

【例 3.1】 Activity 实例

这个例子将通过表 3.2 中的步骤来修改 Hello World! 程序的实现。

表 3.2 Activity 实例步骤

步 骤	描 述
1	打开 Android Studio,创建一个 Android 应用,选择"Empty Activity"选项,在"Name"文本框中输入项目名,在"Package name"文本框中输入包名
2	在"Minimum SDK"下拉列表中选择"API 18: Android 4.3(Jelly Bean)"选项或其他 SDK
3	在"Language"下拉列表中选择"Java"选项
4	选择保存路径,单击"Finish"按钮
5	启动 Genymotion 模拟器,然后在 Android 工程项目中进行如下代码修改

修改工程项目的 java/com.example.activity 目录下的 Java 文件 MainActivity.java,修改代码如下:

```java
package com.example.activity;

import android.app.Activity;
import android.os.Bundle;
import android.util.Log;

public class MainActivity extends Activity {

    String msg = "Android : ";

    /** Called when the activity is first created. */
    @Override
    public void onCreate(Bundle savedInstanceState) {
        super.onCreate(savedInstanceState);
        setContentView(R.layout.activity_main);
        Log.d(msg, "The onCreate() event");
    }

    /** Called when the activity is about to become visible. */
    @Override
    protected void onStart() {
        super.onStart();
        Log.d(msg, "The onStart() event");
    }
```

```java
/** Called when the activity has become visible. */
@Override
protected void onResume() {
    super.onResume();
    Log.d(msg, "The onResume() event");
}

/** Called when another activity is taking focus. */
@Override
protected void onPause() {
    super.onPause();
    Log.d(msg, "The onPause() event");
}

/** Called when the activity is no longer visible. */
@Override
protected void onStop() {
    super.onStop();
    Log.d(msg, "The onStop() event");
}

/** Called just before the activity is destroyed. */
@Override
public void onDestroy() {
    super.onDestroy();
    Log.d(msg, "The onDestroy() event");
}
}
```

Activity 通过 setContentView()方法加载工程项目的 res/layout 目录下的 XML 文件，从而调用 UI 控件。使用 setContentView()方法加载 activity_main.xml 文件的语句如下：

```java
setContentView(R.layout.activity_main);
```

Android 应用程序中可以有一个或多个 Activity，但每个 Activity 都需要在 AndroidManifest.xml 配置文件中进行声明。同时，主 Activity 还需要通过<intent-filter>标签标明其为 MAIN 动作和 LAUNCHER 类，语法如下：

```xml
<?xml version="1.0" encoding="utf-8"?>
<manifest xmlns:android="http://schemas.android.com/apk/res/android"
    package="com.example.activity" >

    <application
        android:allowBackup="true"
        android:icon="@mipmap/ic_launcher"
        android:label="@string/app_name"
        android:supportsRtl="true"
        android:theme="@style/AppTheme" >
        <activity android:name=".MainActivity" >
```

```
            <intent-filter>
                <action android:name="android.intent.action.MAIN" />
                <category android:name="android.intent.category.LAUNCHER" />
            </intent-filter>
        </activity>
    </application>
</manifest>
```

注意： 如果在 AndroidManifest.xml 配置文件中没有任何一个 Activity 标注了 MAIN 动作和 LAUNCHER 类，则用户在手机屏幕上将看不到应用程序的 App 图标。

上述程序在 Genymotion 中的运行效果如图 3.2 所示。

图 3.2　Activity 实例运行效果

返回 Android Studio 界面，如图 3.3 所示，在 logcat 窗格中查看 Activity 调用 Log.d()方法的运行结果。

图 3.3　在 Android Studio 中查看运行结果

在 logcat 窗格中找到 Log.d()方法的运行结果：

```
1406-1406/com.example.activity D/Android :: The onCreate() event
1406-1406/com.example.activity D/Android :: The onStart() event
1406-1406/com.example.activity D/Android :: The onResume() event
```

然后，尝试单击 Genymotion 模拟器中的 Home 按钮 ⌂，在 logcat 窗格中找到 Log.d()方法的运行结果：

```
1406-1406/com.example.activity D/Android :: The onPause() event
1406-1406/com.example.activity D/Android :: The onStop() event
```

再次尝试单击 Genymotion 模拟器中的 Menu 按钮 ≡，在 logcat 窗格中找到 Log.d()方法的运行结果：

```
1406-1406/com.example.activity D/Android :: The onStart() event
1406-1406/com.example.activity D/Android :: The onResume() event
```

最后，尝试单击 Genymotion 模拟器中的 Back 按钮 ↩，在 logcat 窗格中找到 Log.d()方法的运行结果：

```
1406-1406/com.example.activity D/Android :: The onPause() event
1406-1406/com.example.activity D/Android :: The onStop() event
1406-1406/com.example.activity D/Android :: The onDestroy() event
```

第 4 章 Android 应用核心 Intent 和 Filters

在 Android 中，Intent（意图）是一个将要执行的操作的抽象描述。Android 的 3 个核心组件——Activity（活动）、Service（服务）和 Broadcast Receiver（广播接收器）都需要使用 Intent 来激活。Intent 用于相同或者不同应用程序组件间的后期运行时绑定。

对于不同的组件，Android 系统提供了不同的 Intent 启动方法，如表 4.1 所示。

表 4.1 不同组件的 Intent 启动方法

组件类型	启动方法
Activity	startActivity(Intent intent) startActivityForResult(Intent intent, int requestCode)
Service	ComponentName startService(Intent service) Boolean bindService(Intent service, ServiceConnection conn, int flags)
BroadcastReceiver	sendBroadcast(Intent intent) sendBroadcast(Intent intent, String receiverPermission) sendOrderedBroadcast(Intent intent, String receiverPermission, BroadcastReceiver resultReceiver resultReceiver, Handler scheduler, int initialCode, String initialData, Bundle initialExtras) sendOrderedBroadcast(Intent intent, String receiverPermission) sendStickyBroadcast(Intent intent) sendStickyOrderedBroadcst(Intent intent, BroadcastReceiver resultReceiver, Handler scheduler, int initialCode, String initialData, Bundle initialExtras)

例如，现有一个 Activity 需要调用 Android 系统中的浏览器来打开一个 URL。为了达到这个目的，Activity 将发送 ACTION_WEB_SEARCH Intent 到 Android Intent Resolver（意图解析器）中，Android Intent Resolver 通过解析 Activity 列表选择一个最合适的 Activity 来执行这个意图。这里，Android Intent Resolver 将选择 Web Browser Activity（网页浏览器活动）来打开 URL。

下述代码表示 Android Intent Resolver 将打开搜索引擎搜索"tutorialspoint"关键字：

```
String q = "tutorialspoint";
Intent intent = new Intent(Intent.ACTION_WEB_SEARCH );
intent.putExtra(SearchManager.QUERY, q);
startActivity(intent);
```

4.1 Intent 对象的各属性

Intent 对象中包含了接收该 Intent 的组件感兴趣的信息（如执行的操作和操作的数据）和 Android 系统感兴趣的信息（如处理该 Intent 的组件的类别和任何启动目标 Activity 的说明）。Intent 对象大致包含 Component（组件）、Action（动作）、Category（类别）、Data（数据）、Type（类型）、Extra（额外）和 Flag（标记）等属性，下面详细介绍 Intent 对象各属性的作用。

4.1.1 Component（组件）

Intent 对象的 Component 属性需要接收一个 ComponentName（组件名称）属性，ComponentName 属性包含如表 4.2 所示的几个构造器。

表 4.2 ComponentName 属性包含的构造器

构造器	描述
ComponentName(String pkg, String cls)	创建 pkg 所对应的包下的 cls 类所对应的组件
ComponentName(Context pkg, String cls)	创建 pkg 所对应的包下的 cls 类所对应的组件
ComponentName(Context pkg, Class<?> cls)	创建 pkg 所对应的包下的 cls 类所对应的组件

上述构造器的本质为一个构造器，这说明创建一个 ComponentName 需要指定包名和类名（可唯一确定的组件类），这样应用程序就可以根据给定的组件类启动特定的组件。

除此之外，Intent 还包含如表 4.3 所示的 3 个方法。

表 4.3 设置 Intent 启动组件类的 3 个方法

方法	描述
setClass(Context packageContext, Class<?> cls)	设置该 Intent 将要启动的组件的类
setClassName(Context packageContext, String className)	设置该 Intent 将要启动的组件对应的类名
setClassName(String packageName, String className)	设置该 Intent 将要启动的组件对应的类名

4.1.2 Action（动作）

Action 是一个字符串，用来表示要完成的一个抽象动作。而这个动作具体由哪个组件（或许是 Activity、或许是 BroadcastReceiver）来完成，并不是由 Action 这个字符串本身决定的。需要指出的是，一个 Intent 对象最多只能包含一个 Action 属性，Android 定义了一系列 Action 常量，其目标组件包括 Activity 和 BroadcastReceiver 两类。下面分别介绍这两类动作。

1. 启动 Activity 的标准动作

表 4.4 中列出了当前 Intent 类中定义的用于启动 Activity 的标准动作，其中比较常用的是 ACTION_MAIN 和 ACTION_EDIT。

表 4.4 Intent 类中定义的用于启动 Activity 的标准动作

Action 常量	对应字符串	描述
ACTION_MAIN	Android.intent.action.MAIN	应用程序入口
ACTION_VIEW	Android.intent.action.VIEW	显示指定数据
ACTION_ATTACH_DATA	Android.intent.action.ATTACH_DATA	指定某块数据将被附加到其他地方
ACTION_EDIT	Android.intent.action.EDIT	编辑指定数据
ACTION_PICK	Android.intent.action.PICK	从列表中选择某项，并返回所选的数据
ACTION_CHOOSER	Android.intent.action.CHOOSER	显示一个 Activity 选择器
ACTION_GET_CONTENT	Android.intent.action.GET_CONTENT	让用户选择数据，并返回所选数据
ACTION_DIAL	Android.intent.action.DIAL	显示拨号面板
ACTION_CALL	Android.intent.action.CALL	直接向指定用户打电话
ACTION_SEND	Android.intent.action.SEND	向其他人发送数据
ACTION_SENDTO	Android.intent.action.SENDTO	向其他人发送信息
ACTION_ANSWER	Android.intent.action.ANSWER	应答电话

续表

Action 常量	对应字符串	描述
ACTION_INSERT	Android.intent.action.INSERT	插入数据
ACTION_DELETE	Android.intent.action.DELETE	删除数据
ACTION_RUN	Android.intent.action.RUN	运行数据
ACTION_SYNC	Android.intent.action.SYNC	执行数据同步
ACTION_PICK_ACTIVITY	Android.intent.action.ACTIVITY	用于选择 Activity
ACTION_SEARCH	Android.intent.action.SEARCH	执行搜索
ACTION_WEB_SEARCH	Android.intent.action.WEB_SEARCH	执行 Web 搜索
ACTION_FACTORY_TEST	Android.intent.action.FACTORY_TEST	工厂测试的入口

2. 接收广播的标准动作

表 4.5 中列出了当前 Intent 类中定义的用于接收广播的标准动作。

表 4.5 Intent 类中定义的用于接收广播的标准动作

Action 常量	对应字符串	描述
ACTION_TIME_TICK	Android.intent.action.TIME_TICK	每分钟通知一次当前时间
ACTION_TIME_CHANGED	Android.intent.action.TIME_CHANGED	通知时间被修改
ACTION_TIMEZONE_CHANGED	Android.intent.action.TIMEZONE_CHANGED	通知时区被修改
ACTION_BOOT_COMPLETED	Android.intent.action.BOOT_COMPLETED	在系统启动完成后发出一次通知
ACTION_PACKAGE_ADDED	Android.intent.action.PACKAGE_ADDED	通知新应用程序包已经被安装到设备上
ACTION_PACKAGE_CHANGED	Android.intent.action.PACKAGE_CHANGED	通知已经安装的应用程序包已经被修改
ACTION_PACKAGE_REMOVED	Android.intent.action.PACKAGE_REMOVED	通知从设备中删除应用程序包
ACTION_PACKAGE_RESTARTED	Android.intent.action._PACKAGE_RESTARTED	通知用户重启应用程序包,其所有进程都将被关闭
ACTION_PACKAGE_DATA_CLEARED	Android.intent.action.PACKAGE_DATA_CLEARED	通知用户清空应用程序包中的数据
ACTION_UID_REMOVED	Android.intent.action.UID_REMOVED	通知从系统中删除用户 ID 值
ACTION_BATTERY_CHANGED	Android.intent.action.BATTERY_CHANGED	包含充电状态、等级和其他电池信息的广播
ACTION_POWER_CONNECTED	Android.intent.action.POWER_CONNECTED	通知设备已经连接外置电源
ACTION_POWER_DISCONNECTED	Android.intent.action.POWER_DISCONNECTED	通知设备已经移除外置电源
ACTION_SHUTDOWN	Android.intent.action.SHUTDOWN	通知设备已经关闭

注意:表 4.4 和表 4.5 所列出的只是部分较为常用的 Action 常量,关于 Intent 所提供的全部 Action 常量,应参考 Android 官方 API 文档中关于 Intent 的说明。

4.1.3 Category(类别)

Category 也是一个字符串,用于为 Action 增加额外的附加类别信息。程序可调用 Intent 类的 addCategory(string str)方法为 Intent 添加 Category,调用 removeCategory()方法删除上次增加的 Category,调用 getCategory()方法获得当前对象中包含的全部 Category。

在 Intent 对象中可以添加任意多个 Category。与 Action 常量类似,在 Intent 类中也预定义了一些 Category 常量，常用的标准 Category 常量及对应的字符串如表 4.6 所示。

表 4.6 常用的标准 Category 常量及对应的字符串

Category 常量	对应的字符串	简 单 说 明
CATEGORY_DEFAULT	android.intent.category.DEFAULT	默认的 Category
CATEGORY_BROWSABLE	android.intent.category.BROWSABLE	指定该 Activity 能被浏览器安全调用
CATEGORY_TAB	android.intent.category.TAB	指定 Activity 作为 TabActivity 的 Tab 页
CATEGORY_LAUNCHER	android.intent.category.LAUNCHER	Activity 显示在顶级程序列表中
CATEGORY_INFO	android.intent.category.INFO	用于提供包信息
CATEGORY_HOME	android.intent.category.HOME	设置该 Activity 随系统启动而运行
CATEGORY_PREFERENCE	android.intent.category.PREFERENCE	该 Activity 是参数面板
CATEGORY_TEST	android.intent.category.TEST	该 Activity 是一个测试
CATEGORY_CAR_DOCK	android.intent.category.CAR_DOCK	指定手机被插入汽车硬件时运行该 Activity
CATEGORY_DESK_DOCK	android.intent.category.DESK_DOCK	指定手机被插入桌面硬件时运行该 Activity
CATEGORY_CAR_MODE	android.intent.category.CAR_MODE	设置该 Activity 可在车载环境下使用

注意：表 4.6 列出的只是部分较为常用的 Category 常量，关于 Intent 所提供的全部 Category 常量，应参考 Android 官方 API 文档中关于 Intent 的说明。

4.1.4 Data（数据）和 Type（类型）

Data 通常用于向 Action 属性提供操作的数据。Data 接收一个 URI 对象。一个 URI 对象通常通过如下形式的字符串来表示：

```
content://com.android.contacts/contacts/1
tel:123
```

URI 字符串通常采用如下格式：

```
scheme://host:port/path
```

例如，上面给出的 content://com.android.contacts/contacts/1，其中，content 是 scheme 部分；com.android.contacts 是 host 部分；port 部分被省略；/contacts/1 是 path 部分。

Type 用于指定该 Data 所指定的 URI 对应的 MIME 类型。这种 MIME 类型可以是任意自定义的 MIME 类型，只要符合 abc/xyz 格式的字符串即可。

Data 和 Type 的关系比较微妙，这两个属性会相互覆盖，例如：

- 如果为 Intent 先设置 Data，再设置 Type，则 Type 属性将会覆盖 Data 属性。
- 如果为 Intent 先设置 Type，再设置 Data，则 Data 属性将会覆盖 Type 属性。
- 如果希望 Intent 既有 Data 属性，也有 Type 属性，则需要调用 Intent 的 setDataAndType() 方法。

在 AndroidManifest.xml 配置文件中为组件声明 Data、Type 属性时，都需要使用<data.../>标签。<data.../>标签的格式如下：

```
<data android:mimeType=" "
    android:scheme=" "
    android:host=" "
    android:port=" "
    android:path=" "
```

```
android:pathPrefix=" "
android:pathPattern=" " />
```

上面的<data.../>标签支持如下属性。
- mimeType：用于声明该组件所能匹配的 Intent 的 Type 属性。
- scheme：用于声明该组件所能匹配的 Intent 的 Data 属性的 scheme 部分。
- host：用于声明该组件所能匹配的 Intent 的 Data 属性的 host 部分。
- port：用于声明该组件所能匹配的 Intent 的 Data 属性的 port 部分。
- path：用于声明该组件所能匹配的 Intent 的 Data 属性的 path 部分。
- pathPrefix：用于声明该组件所能匹配的 Intent 的 Data 属性的 path 前缀。
- pathPattern：用于声明该组件所能匹配的 Intent 的 Data 属性的 path 字符串模板。

一旦为 Intent 同时指定了 Action 和 Data 属性，Android 就可以根据指定的数据类型来启动特定的应用程序，并对指定数据执行相应的操作。

下面是几个 Action 属性和 Data 属性组合的例子。
- ACTION_VEW content://com.android.contacts/contacts/1：显示标记为 1 的联系人的信息。
- ACTION_EDIT content://com.android.contacts/contacts/1：编辑标记为 1 的联系人的信息。
- ACTION_DIAL content://com.android.contacts/contacts/1：显示向标记为 1 的联系人拨号的界面。
- ACTION_VEW tel:123：显示向指定号码 123 拨号的界面。
- ACTION_DIAL tel:123：显示向指定号码 123 拨号的界面。
- ACTION_VEW content://contacts/people/：显示所有联系人列表的信息。

4.1.5 Extra（额外）

Intent 的 Extra 属性通常用于在多个 Activity 之间进行数据交换。Intent 的 Extra 属性值应该是一个 Bundle 对象，而 Bundle 对象是一个 Map 数据结构对象，它可以存入多组 key-value 对。Extra 可以通过 putExtras()方法和 getExtras()方法设置和读取 Bundle 对象。这样，就可以通过 Intent 在不同 Activity 之间进行数据交换了。

4.1.6 Flag（标记）

Intent 的 Flag 属性用于为该 Intent 添加一些额外的控制标记。Intent 可调用 addFlags()方法为 Intent 添加控制标记。Intent 中常用的 Flag 如表 4.7 所示。

表 4.7 Intent 中常用的 Flag

Flag	描 述
FLAG_ACTIVITY_BROUGHT_TO_FRONT	如果通过该标记启动的 Activity 已经存在，则在下次启动时，只是将该 Activity 调用到前台
FLAG_ACTIVITY_CLEAR_TOP	该标记相当于加载模式中的 singleTask，通过这种 Flag 启动的 Activity 将会把要启动的 Activity 之上的 Activity 全部弹出 Activity 栈
FLAG_ACTIVITY_NEW_TASK	默认的启动标记，该标记用于控制重新创建一个新的 Activity
FLAG_ACTIVITY_NO_ANIMATION	该标记用于控制启动 Activity 时不使用过渡动画
FLAG_ACTIVITY_NO_HISTORY	该标记用于控制被启动的 Activity 不会被保留在 Activity 栈中
FLAG_ACTIVITY_REORDER_TO_FRONT	该标记用于控制如果当前已有该 Activity，则直接将该 Activity 调用到前台
FLAG_ACTIVITY_SINGLE_TOP	该标记相当于加载模式中的 singleTop 模式

4.2　Intent 的类型

在 Android 中，Intent 分为显式 Intent 和隐式 Intent 两种类型，如图 4.1 所示。

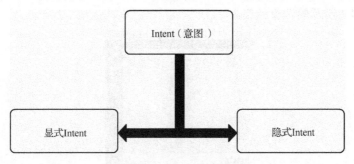

图 4.1　Intent 的类型

4.2.1　显式 Intent

显式 Intent 是指定了 Component 属性的 Intent，它已经明确了将启动哪个组件。例如，从一个 Activity 中通过单击按钮启动另一个 Activity，运行效果如图 4.2 所示。

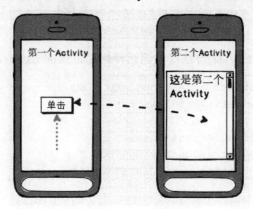

图 4.2　显式 Intent 举例

通过如下代码即可实现：

```
// Explicit Intent by specifying its class name
Intent i = new Intent(FirstActivity.this, SecondActivity.class);
i.putExtra("Key1", "ABC");
i.putExtra("Key2", "123");
// Starts TargetActivity
startActivity(i);
```

4.2.2　隐式 Intent

隐式 Intent 是没有指定 Component 属性的 Intent。由于隐式 Intent 没有明确指定要启动哪个组件，因此应用程序将会根据 Intent 指定的规则启动符合条件的组件，但具体启动哪个组件则不确定（由系统自动选择适合的组件来执行 Intent）。例如，运行下面的代码：

```
Intent read1=new Intent();
read1.setAction(android.content.Intent.ACTION_VIEW);
read1.setData(ContactsContract.Contacts.CONTENT_URI);
startActivity(read1);
```

将得到如图 4.3 所示的运行效果。

图 4.3 隐式 Intent 举例

目标组件可以通过 getExtras()方法接收源组件传递过来的 Extra 信息，代码如下：

```
// Get bundle object at appropriate place in your code
Bundle extras = getIntent().getExtras();

// Extract data using passed keys
String value1 = extras.getString("Key1");
String value2 = extras.getString("Key2");
```

【例 4.1】 Intent 实例

这个例子将展示 Intent 在 Android 中的应用。Intent 实例 1 步骤如表 4.8 所示。

表 4.8 Intent 实例 1 步骤

步骤	描述
1	打开 Android Studio，创建一个 Android 应用，选择"Empty Activity"选项，在"Name"文本框中输入项目名，在"Package name"文本框中输入包名
2	在"Minimum SDK"下拉列表中选择"API 18: Android 4.3（Jelly Bean）"选项或其他 SDK
3	在"Language"下拉列表中选择"Java"选项
4	选择保存路径，单击"Finish"按钮
5	在工程项目中找到 ras/layout 目录中的 activity_main.xml 文件，并在其中添加一个 TextView、一个 ImageButton 和两个 Button（START BROWSER 和 START PHONE）控件，按如图 4.4 所示的运行效果进行摆放
6	将工程项目的 res/mipmap 目录下的图片文件 ic_launcher.png 复制、粘贴到 res/drawable 目录下
7	修改 java/com.example.intent1 目录下的 Java 文件 MainActivity.java 的代码，为 START BROWSER 和 START PHONE 两个按钮添加单击事件监听器
8	启动 Genymotion 模拟器，然后在 Android 工程项目中进行如下代码修改

修改新建工程项目的 res/layout 目录下的布局文件 activity_main.xml，修改代码如下：

```xml
<?xml version="1.0" encoding="utf-8"?>
<RelativeLayout xmlns:android="http://schemas.android.com/apk/res/android"
    xmlns:tools="http://schemas.android.com/tools"
    android:layout_width="match_parent"
    android:layout_height="match_parent"
    android:paddingLeft="@dimen/activity_horizontal_margin"
    android:paddingRight="@dimen/activity_horizontal_margin"
    android:paddingTop="@dimen/activity_vertical_margin"
    android:paddingBottom="@dimen/activity_vertical_margin"
    tools:context=".MainActivity">

    <TextView
        android:id="@+id/textView1"
        android:layout_width="wrap_content"
        android:layout_height="wrap_content"
        android:text="Intent Example"
        android:layout_alignParentTop="true"
        android:layout_centerHorizontal="true"
        android:textSize="30dp" />

    <ImageButton
        android:layout_width="wrap_content"
        android:layout_height="wrap_content"
        android:id="@+id/imageButton"
        android:src="@drawable/ic_launcher"
        android:layout_centerHorizontal="true"
        android:layout_below="@+id/textView1" />

    <Button
        android:layout_width="wrap_content"
        android:layout_height="wrap_content"
        android:text="START BROWSER"
        android:id="@+id/button"
        android:layout_alignLeft="@+id/imageButton"
        android:layout_alignStart="@+id/imageButton"
        android:layout_below="@+id/imageButton"
        android:layout_alignEnd="@+id/imageButton" />

    <Button
        android:layout_width="wrap_content"
        android:layout_height="wrap_content"
        android:text="START PHONE"
        android:id="@+id/button2"
        android:layout_below="@+id/button"
        android:layout_alignLeft="@+id/button"
        android:layout_alignStart="@+id/button"
        android:layout_alignEnd="@+id/imageButton" />
```

```
</RelativeLayout>
```

修改工程项目的 java/com.example.intent1 目录下的 Java 文件 MainActivity.java，修改代码如下：

```java
package com.example.intent1;

import android.app.Activity;
import android.content.Intent;
import android.net.Uri;
import android.os.Bundle;
import android.view.View;
import android.widget.Button;

public class MainActivity extends Activity {
    Button b1,b2;
    @Override
    protected void onCreate(Bundle savedInstanceState) {
        super.onCreate(savedInstanceState);
        setContentView(R.layout.activity_main);
        b1=(Button)findViewById(R.id.button);
        b1.setOnClickListener(new View.OnClickListener() {

            @Override
            public void onClick(View v) {
                Intent i = new Intent(android.content.Intent.ACTION_VIEW,
                        Uri.parse("http://www.baidu.com"));
                startActivity(i);
            }
        });

        b2=(Button)findViewById(R.id.button2);
        b2.setOnClickListener(new View.OnClickListener() {
            @Override
            public void onClick(View v) {
                Intent i = new Intent(android.content.Intent.ACTION_VIEW,
                        Uri.parse("tel:18611111186"));
                startActivity(i);
            }
        });
    }
}
```

程序在 Genymotion 中的运行效果如图 4.4 所示。

单击"START BROWSER"按钮，运行效果如图 4.5 所示。

单击"START PHONE"按钮，运行效果如图 4.6 所示。

第 4 章 Android 应用核心 Intent 和 Filters

图 4.4 Intent 实例 1
运行效果（1）

图 4.5 Intent 实例 1
运行效果（2）

图 4.6 Intent 实例 1
运行效果（3）

4.3 Intent Filters（意图过滤器）

Activity、Service 和 BroadcastReceiver 可以通过定义多个 Intent 过滤器来通知系统它们可以处理哪些隐式 Intent。每个过滤器都会描述组件的一种能力及该组件可以接收的一组 Intent。实际上，过滤器接收需要类型的 Intent，拒绝不需要类型的 Intent，这仅限于隐式 Intent。对于显式 Intent 而言，无论内容如何，都可以发送给目标，而过滤器不干预。

过滤器使用<intent-filter.../>标签在配置文件 AndroidManifest.xml 中进行注册。该标准是<activity.../>标签的子标签，<activity.../>标签用于为应用程序配置 Activity，<activity.../>的<intent-filter.../>标签则用于配置该 Activity 所能"响应"的 Intent。

<intent-filter.../>标签中通常包含如下子标签。

- 1～N 个<action.../>子标签。
- 0～N 个<category.../>子标签。
- 0～1 个<data.../>子标签。

以下面这个<activity.../>的标签为例：

```xml
<activity android:name=".CustomActivity"
    android:label="@string/app_name">
    <intent-filter>
        <action android:name="android.intent.action.VIEW" />
        <action android:name="com.example.intentdemo.LAUNCH" />
        <category android:name="android.intent.category.DEFAULT" />
        <data android:scheme="http" />
    </intent-filter>
</activity>
```

根据上述代码，这个 Activity 过滤器有两个<action.../>子标签，表示其他 Activity 调用该 Activity 时要么使用 android.intent.action.VIEW 动作，要么使用 com.example.intentdemo.LAUNCH 动作。

一个<category.../>子标签表示类别为 android.intent.category.DEFAULT。

一个<data.../>子标签表示需要 Activity 启动的数据类型为 http://。

那么，如果一个 Intent 可以通过多个 Activity 或 Service 组件来执行，系统到底会调用哪一个 Activity 或 Service 组件来执行这个 Intent 呢？

Android 系统会通过如下的测试来决定调用哪一个组件。

- <intent-filter.../>标签可以包含 1 个或多个<action.../>子标签，系统会选择其中最匹配的动作来调用 Activity。但<action.../>子标签不能采用默认设置，即过滤器必须包含至少一个<action.../>子标签，否则会阻塞所有 Intent。
- <intent-filter.../>标签可以包含 0 个、1 个或多个<category.../>子标签。0 个表示 Intent 总是可以通过该项测试；1 个或多个表示 Intent 类别必须和每一种类别相匹配才能通过测试。需要注意的是，Android 默认所有通过 startActivity()方法传递的隐式 Intent 都包含一个默认类型 android.intent.category.DEFAULT（CATEGORY_DEFAULT 常量），因此，接收隐式 Intent 的 Activity 必须在过滤器中包含 android.intent.category.DEFAULT 以与之相匹配。
- <intent-filter.../>标签可以包含 0 个或 1 个< data .../>子标签。每个< data .../>子标签可以指定 URI 和数据类型（MIME 媒体类型）。URI 可以分成 scheme、host、port 和 path 几个独立的部分。Intent 对象的 URI 和数据类型两部分都要和过滤器的 URI 和数据类型相匹配才能通过测试，匹配规则如表 4.9 所示。

表 4.9 匹配规则

Intent 对象		过 滤 器		通 过 条 件
URI	数 据 类 型	URI	数 据 类 型	
未指定	未指定	未指定	未指定	无条件通过
指定	未指定	指定	未指定	两个 URI 匹配
未指定	指定	未指定	指定	两个数据类型匹配
指定	指定	指定	指定	URI 和数据类型匹配

【例 4.2】 Intent 使用过滤器实例

这个案例将简单修改上一个 Intent 实例，我们将会看到一个 Intent 通过调用两个 Activity 来让用户做出选择的例子、一个用过滤器测试 Intent 调用 Activity 的例子和一个因 Intent 没有适合的 Activity 而调用出错的例子。Intent 实例 2 步骤如表 4.10 所示。

表 4.10 Intent 实例 2 步骤

步 骤	描 述
1	打开 Android Studio，创建一个 Android 应用，选择"Empty Activity"选项，在"Name"文本框中输入项目名，在"Package name"文本框中输入包名
2	在"Minimum SDK"下拉列表中选择"API 18: Android 4.3（Jelly Bean）"选项或其他 SDK
3	在"Language"下拉列表中选择"Java"选项
4	选择保存路径，单击"Finish"按钮
5	将工程项目的 res/mipmap 目录下的图片文件 ic_launcher.png 复制、粘贴到 res/drawable 目录下

续表

步骤	描 述
6	在工程项目中找到 ras/layout 目录中的 activity_main.xml 文件,并在其中添加一个 TextView、一个 ImageButton 和三个 Button(Start browsing with view action、Start browsing with launch action 和 Exceptional condition)控件,按如图 4.7 所示的运行效果进行摆放
7	修改 java/com.example.intent2 目录下的 Java 文件 MainActivity.java 的代码,为 Start browsing with view action、Start browsing with launch action 和 Exceptional condition 三个按钮添加单击事件监听器
8	右击 java/com.example.intent2 目录,在弹出的快捷菜单中选择 "new"→"Activity"→"Empty Activity" 命令,新建一个 Activity 文件并将其命名为 CustomActivity.java, 按下面的代码修改文件
9	在工程项目中找到 ras/layout 目录中随 CustomActivity 新生成的 activity_custom.xml 文件,并在其中添加一个 TextView 控件,用于显示 Intent 传来的值
10	按下面的代码修改 manifests 目录下的配置文件 AndroidManifest.xml
11	启动 Genymotion 模拟器,然后在 Android 工程项目中进行如下代码修改

修改新建工程项目的 res/layout 目录下的布局文件 activity_main.xml,修改代码如下:

```xml
<?xml version="1.0" encoding="utf-8"?>
<RelativeLayout xmlns:android="http://schemas.android.com/
                                apk/res/android"
    xmlns:tools="http://schemas.android.com/tools"
    android:layout_width="match_parent"
    android:layout_height="match_parent"
    android:paddingLeft="@dimen/activity_horizontal_margin"
    android:paddingRight="@dimen/activity_horizontal_margin"
    android:paddingTop="@dimen/activity_vertical_margin"
    android:paddingBottom="@dimen/activity_vertical_margin"
    tools:context=".MainActivity">

    <TextView
        android:id="@+id/textView1"
        android:layout_width="wrap_content"
        android:layout_height="wrap_content"
        android:text="Intent Example"
        android:layout_alignParentTop="true"
        android:layout_centerHorizontal="true"
        android:textSize="30dp" />

    <ImageButton
        android:layout_width="wrap_content"
        android:layout_height="wrap_content"
        android:id="@+id/imageButton"
        android:src="@drawable/ic_launcher"
        android:layout_centerHorizontal="true"
        android:layout_below="@+id/textView1" />

    <Button
        android:layout_width="wrap_content"
        android:layout_height="wrap_content"
        android:text="Start browsing with view action"
```

```xml
        android:id="@+id/button"
        android:layout_alignLeft="@+id/imageButton"
        android:layout_alignStart="@+id/imageButton"
        android:layout_below="@+id/imageButton"
        android:layout_alignEnd="@+id/imageButton" />

    <Button
        android:layout_width="wrap_content"
        android:layout_height="wrap_content"
        android:text="Start browsing with launch action"
        android:id="@+id/button2"
        android:layout_below="@+id/button"
        android:layout_alignLeft="@+id/button"
        android:layout_alignStart="@+id/button"
        android:layout_alignEnd="@+id/button" />

    <Button
        android:layout_width="wrap_content"
        android:layout_height="wrap_content"
        android:text="Exceptional condition"
        android:id="@+id/button3"
        android:layout_below="@+id/button2"
        android:layout_alignLeft="@+id/button2"
        android:layout_alignStart="@+id/button2"
        android:layout_alignEnd="@+id/button2" />
</RelativeLayout>
```

修改工程项目的 java/com.example.intent2 目录下的 Java 文件 MainActivity.java，修改代码如下：

```java
package com.example.intent2;

import android.app.Activity;
import android.content.Intent;
import android.net.Uri;
import android.os.Bundle;
import android.view.View;
import android.widget.Button;

public class MainActivity extends Activity {
    Button b1,b2,b3;
    @Override
    protected void onCreate(Bundle savedInstanceState) {
        super.onCreate(savedInstanceState);
        setContentView(R.layout.activity_main);
        b1=(Button)findViewById(R.id.button);

        b1.setOnClickListener(new View.OnClickListener() {

            @Override
```

```java
            public void onClick(View v) {
                Intent i=new Intent(android.content.Intent.ACTION_VIEW,Uri.parse
                    ("http://www.baidu.com"));
                startActivity(i);
            }
        });

        b2=(Button)findViewById(R.id.button2);
        b2.setOnClickListener(new View.OnClickListener() {

            @Override
            public void onClick(View v) {
                Intent i=new Intent("com.example.intent2.LAUNCH", Uri.parse
                    ("http://www.baidu.com"));
                startActivity(i);
            }
        });

        b3=(Button)findViewById(R.id.button3);
        b3.setOnClickListener(new View.OnClickListener() {

            @Override
            public void onClick(View v) {
                Intent i = new Intent("com.example.intent2.LAUNCH",
                    Uri.parse("https://www.baidu.com"));
                startActivity(i);
            }
        });
    }
}
```

修改工程项目的 res/layout 目录下的布局文件 activity_custom.xml，修改代码如下：

```xml
<?xml version="1.0" encoding="utf-8"?>
<RelativeLayout xmlns:android="http://schemas.android.com/
                                apk/res/android"
    xmlns:tools="http://schemas.android.com/tools"
    android:layout_width="match_parent"
    android:layout_height="match_parent"
    android:paddingLeft="@dimen/activity_horizontal_margin"
    android:paddingRight="@dimen/activity_horizontal_margin"
    android:paddingTop="@dimen/activity_vertical_margin"
    android:paddingBottom="@dimen/activity_vertical_margin"
    tools:context="com.example.intent2.CustomActivity">
    <TextView android:id="@+id/show_data"
        android:layout_width="fill_parent"
        android:layout_height="400dp" />
</RelativeLayout>
```

修改工程项目的 java/com.example.intent2 目录下的 Java 文件 CustomActivity.java，修改代码如下：

```java
package com.example.intent2;

import android.net.Uri;
import android.os.Bundle;
import android.support.v7.app.AppCompatActivity;
import android.widget.TextView;

public class CustomActivity extends AppCompatActivity {

    @Override
    protected void onCreate(Bundle savedInstanceState) {
        super.onCreate(savedInstanceState);
        setContentView(R.layout.activity_custom);
        TextView label = (TextView) findViewById(R.id.show_data);
        Uri url = getIntent().getData();
        label.setText(url.toString());
    }
}
```

修改工程项目的 manifests 目录下的配置文件 AndroidManifest.xml,修改代码如下:

```xml
<?xml version="1.0" encoding="utf-8"?>
<manifest xmlns:android="http://schemas.android.com/apk/res/android"
    package="com.example.intent2" >

    <application
        android:allowBackup="true"
        android:icon="@mipmap/ic_launcher"
        android:label="@string/app_name"
        android:supportsRtl="true"
        android:theme="@style/AppTheme" >
        <activity android:name=".MainActivity" >
            <intent-filter>
                <action android:name="android.intent.action.MAIN" />

                <category android:name="android.intent.category.LAUNCHER" />
            </intent-filter>
        </activity>

        <activity android:name=".CustomActivity" >
            <intent-filter>
                <action android:name="android.intent.action.VIEW" />
                <action android:name="com.example.intent2.LAUNCH" />
                <category android:name="android.intent.category.DEFAULT" />
                <data android:scheme="http" />
            </intent-filter>
        </activity>
    </application>

</manifest>
```

程序在 Genymotion 中的运行效果如图 4.7 所示。

单击"Start browsing with view action"按钮，由于我们已经在 CustomActivity 的过滤器中定义了如下两个动作标签：<action android:name="android.intent.action.VIEW" />标签表示系统会调用网页浏览器来响应 Intent；<action android:name="com.example.intent2.LAUNCH" />标签则表示用 CustomActivity 来响应 Intent，两者都是可行的。因此，将会打开如图 4.8 所示的选择对话框，让用户选择一个组件来响应该 Intent。

图 4.7 Intent 实例 2 运行效果

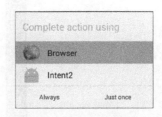

图 4.8 选择对话框

如果用户选择了"Browser"选项，则 Android 将启动网页浏览器并打开网址"http://www.baidu.com"，如图 4.9 所示。

如果用户选择了"Intent2"选项，则 Android 将调用 CustomActivity 组件并把 Intent 值"http://www.baidu.com"传递给其中的文本框，如图 4.10 所示。

图 4.9 调用系统组件响应 Intent 请求

图 4.10 调用 CustomActivity 组件响应 Intent 请求

单击模拟器中的 Back 按钮 ⇦，返回实例主界面。单击"Start browsing with launch action"按钮，这时 Android 会启用过滤器对组件进行测试，请求 Intent 与 CustomActivity 组件过滤器相匹配，所以系统直接调用 CustomActivity 组件并把 Intent 值"http://www.baidu.com"传递给其中的文本框，运行效果如图 4.11 所示。

再次单击模拟器中的 Back 按钮 ⇦，返回实例主界面。单击"Exceptional condition"按钮，由于我们把 Intent 中的请求数据 http 修改成了 https，Android 系统不能找到与 Intent 请求相匹配的组件，因此系统将提示错误，并退出应用程序，如图 4.12 所示。

图 4.11 运行效果

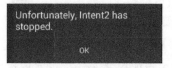

图 4.12 系统提示错误

第 5 章 Android 事件处理

无论是桌面应用程序还是手机应用程序，面对最多的都是用户，经常需要处理的就是用户动作，也就是需要响应用户动作，这种响应用户动作的机制就是事件处理。

5.1 Android 事件处理概述

下面介绍 Android 事件处理的 3 个基本概念。
- 事件监听器：事件监听器就是包含一个回调方法的 View（视图）类接口，它负责监听 UI 上的 View（视图）类控件发生的各种事件，并对事件做出相应的响应。
- 事件监听器注册：事件监听器注册是指将事件监听器与一个事件处理程序绑定，当事件发生时，事件监听器侦测到事件后交由与其绑定（注册）的事件处理程序进行响应处理。
- 事件处理程序：当事件发生时，事件监听器将调用与其绑定（注册）的事件处理程序以处理事件。事件处理程序中包含处理事件的实例方法。

常见的事件处理程序与事件监听器如表 5.1 所示。

表 5.1 常见的事件处理程序与事件监听器

事件处理程序	事件监听器
onClick()	OnClickListener()： 当用户单击、触摸或把焦点放到按钮、文本或图片等控件上时，它将调用 onClick() 事件处理程序来响应该事件
onLongClick()	OnLongClickListener()： 当用户单击、触摸或把焦点放到按钮、文本或图片等控件上的时间超过 1s 时，它将调用 onLongClick() 事件处理程序来响应该事件
onFocusChange()	OnFocusChangeListener()： 当控件失去焦点时，它将调用 onFocusChange() 事件处理程序来响应该事件
onTouch()	OnTouchListener()： 当用户按下硬件按键、松开硬件按键或在屏幕上做任何手势时，它将调用 onTouch() 事件处理程序来响应该事件
onMenuItemClick()	OnMenuItemClickListener()： 当用户选中一个菜单项时，它将调用 onMenuItemClick() 事件处理程序来响应该事件

注意：在 Android 中还有很多其他的事件监听器，如 OnHoverListener()、OnDragListener() 等，但这里并没有列出来。当开发一个复杂应用程序时，请查看 Android 官方 API 文档。

5.2 事件监听器的注册方法

注册事件监听器就是将事件监听器与一个事件处理程序绑定。事件监听器的注册方法有很多种，但常用的方法有以下 3 种。

- 使用匿名内部类。
- 使用 Activity 类执行 Listener（监听器）接口。
- 在布局文件 activity_main.xml 中直接指定事件处理程序（方法）。

下面举例说明以上 3 种方法。

【例 5.1】 基于匿名内部类的事件监听器注册

这个例子将通过几个简单的步骤展示如何使用 OnClickListener 类注册事件监听器并捕获单击事件。基于匿名内部类的事件监听器注册实例步骤如表 5.2 所示。

表 5.2 基于匿名内部类的事件监听器注册实例步骤

步骤	描述
1	打开 Android Studio，创建一个 Android 应用，选择"Empty Activity"选项，在"Name"文本框中输入项目名，在"Package name"文本框中输入包名
2	在"Minimum SDK"下拉列表中选择"API 18: Android 4.3（Jelly Bean）"选项或其他 SDK
3	在"Language"下拉列表中选择"Java"选项
4	选择保存路径，单击"Finish"按钮
5	将工程项目的 res/mipmap 目录下的图片文件 ic_launcher.png 复制、粘贴到 res/drawable 目录下
6	在工程项目中找到 ras/layout 目录中的 activity_main.xml 文件，并在其中添加一个 TextView、一个 ImageButton 和两个 Button（SMALL FONT 和 LARGE FONT）控件，按如图 5.1 所示的运行效果进行摆放
7	修改 java/com.example.eventdemo1 目录下的 Java 文件 MainActivity.java 的代码，为 SMALL FONT 和 LARGE FONT 两个按钮添加单击事件监听器

修改新建工程项目的 res/layout 目录下的布局文件 activity_main.xml，修改代码如下：

```xml
<?xml version="1.0" encoding="utf-8"?>
<RelativeLayout xmlns:android="http://schemas.android.com/
                                apk/res/android"
    xmlns:tools="http://schemas.android.com/tools"
    android:layout_width="match_parent"
    android:layout_height="match_parent"
    android:paddingLeft="@dimen/activity_horizontal_margin"
    android:paddingRight="@dimen/activity_horizontal_margin"
    android:paddingTop="@dimen/activity_vertical_margin"
    android:paddingBottom="@dimen/activity_vertical_margin"
    tools:context=".MainActivity">

    <TextView
        android:id="@+id/textView1"
        android:layout_width="wrap_content"
        android:layout_height="wrap_content"
        android:text="Event Handling"
        android:layout_alignParentTop="true"
        android:layout_centerHorizontal="true"
        android:textSize="30dp"/>

    <ImageButton
        android:layout_width="wrap_content"
        android:layout_height="wrap_content"
```

```xml
        android:id="@+id/imageButton"
        android:src="@drawable/ic_launcher"
        android:layout_centerVertical="true"
        android:layout_centerHorizontal="true" />

    <Button
        android:layout_width="wrap_content"
        android:layout_height="wrap_content"
        android:text="SMALL FONT"
        android:id="@+id/button"
        android:layout_below="@+id/imageButton"
        android:layout_alignEnd="@+id/textView1"
        android:layout_alignStart="@+id/textView1" />

    <Button
        android:layout_width="wrap_content"
        android:layout_height="wrap_content"
        android:text="LARGE FONT"
        android:id="@+id/button2"
        android:layout_below="@+id/button"
        android:layout_alignRight="@+id/button"
        android:layout_alignEnd="@+id/button"
        android:layout_alignStart="@+id/button" />

    <TextView
        android:layout_width="wrap_content"
        android:layout_height="wrap_content"
        android:text="Hello World!"
        android:id="@+id/textView"
        android:layout_below="@+id/button2"
        android:layout_centerHorizontal="true"
        android:textSize="25dp" />
</RelativeLayout>
```

修改工程项目的 java/com.example.eventdemo1 目录下的 Java 文件 MainActivity.java，修改代码如下：

```java
package com.example.eventdemo1;

import android.os.Bundle;
import android.support.v7.app.AppCompatActivity;
import android.view.View;
import android.widget.Button;
import android.widget.TextView;

public class MainActivity extends AppCompatActivity {

    @Override
    protected void onCreate(Bundle savedInstanceState) {
```

```java
        super.onCreate(savedInstanceState);
        setContentView(R.layout.activity_main);
        // --- find both the buttons---
        Button sButton = (Button) findViewById(R.id.button);
        Button lButton = (Button) findViewById(R.id.button2);

        // -- register click event with first button ---
        sButton.setOnClickListener(new View.OnClickListener() {
            public void onClick(View v) {
                // --- find the text view --
                TextView txtView = (TextView) findViewById(R.id.textView);
                // -- change text size --
                txtView.setTextSize(14);
            }
        });

        // -- register click event with second button ---
        lButton.setOnClickListener(new View.OnClickListener() {
            public void onClick(View v) {
                // --- find the text view --
                TextView txtView = (TextView) findViewById(R.id.textView);
                // -- change text size --
                txtView.setTextSize(24);
            }
        });

    }
}
```

程序在 Genymotion 中的运行效果如图 5.1 所示。

图 5.1　基于匿名内部类的事件监听器注册实例运行效果

第 5 章 Android 事件处理

【例 5.2】 使用 Activity 类执行 Listener 接口来注册事件监听器

这个例子将通过几个简单的步骤展示如何使用 Activity 类执行 Listener 接口来注册事件监听器并捕获单击事件。使用 Activity 类执行 Listener 接口来注册事件监听器实例步骤如表 5.3 所示。

表 5.3 使用 Activity 类执行 Listener 接口来注册事件监听器实例步骤

步骤	描述
1	打开 Android Studio，创建一个 Android 应用，选择"Empty Activity"选项，在"Name"文本框中输入项目名，在"Package name"文本框中输入包名
2	在"Minimum SDK"下拉列表中选择"API 18: Android 4.3（Jelly Bean）"选项或其他 SDK
3	在"Language"下拉列表中选择"Java"选项
4	选择保存路径，单击"Finish"按钮
5	将工程项目的 res/mipmap 目录下的图片文件 ic_launcher.png 复制、粘贴到 res/drawable 目录下
6	在工程项目中找到 ras/layout 目录中的 activity_main.xml 文件，并在其中添加一个 TextView、一个 ImageButton 和两个 Button（Small font 和 Large Font）控件，按如图 5.2 所示的运行效果进行摆放
7	修改 java/com.example.eventdemo2 目录下的 Java 文件 MainActivity.java 的代码

修改新建工程项目的 res/layout 目录下的布局文件 activity_main.xml，修改代码如下：

```xml
<?xml version="1.0" encoding="utf-8"?>
<RelativeLayout xmlns:android="http://schemas.android.com/
                                apk/res/android"
    xmlns:tools="http://schemas.android.com/tools"
    android:layout_width="match_parent"
    android:layout_height="match_parent"
    android:paddingLeft="@dimen/activity_horizontal_margin"
    android:paddingRight="@dimen/activity_horizontal_margin"
    android:paddingTop="@dimen/activity_vertical_margin"
    android:paddingBottom="@dimen/activity_vertical_margin"
    tools:context=".MainActivity">

    <TextView
        android:id="@+id/textView1"
        android:layout_width="wrap_content"
        android:layout_height="wrap_content"
        android:text="Event Handling"
        android:layout_alignParentTop="true"
        android:layout_centerHorizontal="true"
        android:textSize="30dp" />

    <ImageButton
        android:layout_width="wrap_content"
        android:layout_height="wrap_content"
        android:id="@+id/imageButton"
        android:src="@drawable/ic_launcher"
        android:layout_centerVertical="true"
        android:layout_centerHorizontal="true" />

    <Button
```

```xml
        android:layout_width="wrap_content"
        android:layout_height="wrap_content"
        android:text="Small font"
        android:id="@+id/button"
        android:layout_below="@+id/imageButton"
        android:layout_alignEnd="@+id/imageButton"
        android:layout_alignStart="@+id/textView1" />

    <Button
        android:layout_width="wrap_content"
        android:layout_height="wrap_content"
        android:text="Large Font"
        android:id="@+id/button2"
        android:layout_below="@+id/button"
        android:layout_alignRight="@+id/button"
        android:layout_alignEnd="@+id/button"
        android:layout_alignStart="@+id/button" />

    <TextView
        android:layout_width="wrap_content"
        android:layout_height="wrap_content"
        android:text="Hello World!"
        android:id="@+id/textView"
        android:layout_below="@+id/button2"
        android:layout_centerHorizontal="true"
        android:textSize="25dp" />
</RelativeLayout>
```

修改工程项目的 java/com.example.eventdemo2 目录下的 Java 文件 MainActivity.java,修改代码如下:

```java
package com.example.eventdemo2;

import android.app.Activity;
import android.os.Bundle;
import android.view.View;
import android.widget.Button;
import android.widget.TextView;

public class MainActivity extends Activity implements View.OnClickListener {

    @Override
    protected void onCreate(Bundle savedInstanceState) {
        super.onCreate(savedInstanceState);
        setContentView(R.layout.activity_main);
        // --- find both the buttons---
        Button sButton = (Button) findViewById(R.id.button);
        Button lButton = (Button) findViewById(R.id.button2);

        // -- register click event with first button ---
        sButton.setOnClickListener(this);
```

```
            // -- register click event with second button ---
            lButton.setOnClickListener(this);
    }

    // --- Implement the OnClickListener callback
    public void onClick(View v) {
        if (v.getId() == R.id.button) {
            // --- find the text view --
            TextView txtView = (TextView) findViewById(R.id.textView);
            // -- change text size --
            txtView.setTextSize(14);
            return;
        }
        if (v.getId() == R.id.button2) {
            // --- find the text view --
            TextView txtView = (TextView) findViewById(R.id.textView);
            // -- change text size --
            txtView.setTextSize(24);
            return;
        }
    }
}
```

程序在 Genymotion 中的运行效果如图 5.2 所示。

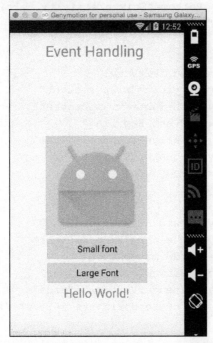

图 5.2　使用 Activity 类执行 Listener 接口来注册事件监听器实例运行效果

【例 5.3】　使用布局文件 activity_main.xml 注册事件监听器

这个例子将通过几个简单的步骤展示如何使用布局文件 activity_main.xml 注册事件监听器并捕获单击事件。使用布局文件 activity_main.xml 注册事件监听器实例步骤如表 5.4 所示。

表 5.4 使用布局文件 activity_main.xml 注册事件监听器实例步骤

步骤	描述
1	打开 Android Studio，创建一个 Android 应用，选择"Empty Activity"选项，在"Name"文本框中输入项目名，在"Package name"文本框中输入包名
2	在"Minimum SDK"下拉列表中选择"API 18: Android 4.3（Jelly Bean）"选项或其他 SDK
3	在"Language"下拉列表中选择"Java"选项
4	选择保存路径，单击"Finish"按钮
5	将工程项目的 res/mipmap 目录下的 ic_launcher.png 图片文件复制、粘贴到 res/drawable 目录下
6	在工程项目中找到 ras/layout 目录中的 activity_main.xml 文件，并在其中添加一个 TextView、一个 ImageButton 和两个 Button（SMALL FONT 和 LARGE FONT）控件，按如图 5.3 所示的运行效果进行摆放，并按如下代码修改
7	按如下代码修改 java/com.example.eventdemo3 目录下的 Java 文件 MainActivity.java

修改新建工程项目的 res/layout 目录下的布局文件 activity_main.xml，修改代码如下：

```xml
<?xml version="1.0" encoding="utf-8"?>
<RelativeLayout xmlns:android="http://schemas.android.com/
                                apk/res/android"
    xmlns:tools="http://schemas.android.com/tools"
    android:layout_width="match_parent"
    android:layout_height="match_parent"
    android:paddingLeft="@dimen/activity_horizontal_margin"
    android:paddingRight="@dimen/activity_horizontal_margin"
    android:paddingTop="@dimen/activity_vertical_margin"
    android:paddingBottom="@dimen/activity_vertical_margin"
    tools:context=".MainActivity">

    <TextView
        android:id="@+id/textView1"
        android:layout_width="wrap_content"
        android:layout_height="wrap_content"
        android:text="Event Handling"
        android:layout_alignParentTop="true"
        android:layout_centerHorizontal="true"
        android:textSize="30dp" />

    <ImageButton
        android:layout_width="wrap_content"
        android:layout_height="wrap_content"
        android:id="@+id/imageButton"
        android:src="@drawable/ic_launcher"
        android:layout_centerVertical="true"
        android:layout_centerHorizontal="true" />

    <Button
        android:layout_width="wrap_content"
        android:layout_height="wrap_content"
        android:text="SMALL FONT"
        android:id="@+id/button"
        android:onClick="doSmall"
```

```xml
        android:layout_below="@+id/imageButton"
        android:layout_alignEnd="@+id/imageButton"
        android:layout_alignStart="@+id/textView1" />

    <Button
        android:layout_width="wrap_content"
        android:layout_height="wrap_content"
        android:text="LARGE FONT"
        android:id="@+id/button2"
        android:onClick="doLarge"
        android:layout_below="@+id/button"
        android:layout_alignRight="@+id/button"
        android:layout_alignEnd="@+id/button"
        android:layout_alignStart="@+id/button" />

    <TextView
        android:layout_width="wrap_content"
        android:layout_height="wrap_content"
        android:text="Hello World!"
        android:id="@+id/textView"
        android:layout_below="@+id/button2"
        android:layout_centerHorizontal="true"
        android:textSize="25dp" />
</RelativeLayout>
```

修改工程项目的 java/com.example.eventdemo3 目录下的 Java 文件 MainActivity.java，修改代码如下：

```java
package com.example.eventdemo3;

import android.os.Bundle;
import android.support.v7.app.AppCompatActivity;
import android.view.View;
import android.widget.TextView;

public class MainActivity extends AppCompatActivity {

    @Override
    protected void onCreate(Bundle savedInstanceState) {
        super.onCreate(savedInstanceState);
        setContentView(R.layout.activity_main);
    }
    // --- Implement the event handler for the first button.
    public void doSmall(View v) {
        // --- find the text view --
        TextView txtView = (TextView) findViewById(R.id.textView);
        // -- change text size --
        txtView.setTextSize(14);
        return;
    }
```

```
        // --- Implement the event handler for the second button.
        public void doLarge(View v)  {
            // --- find the text view --
            TextView txtView = (TextView) findViewById(R.id.textView);
            // -- change text size --
            txtView.setTextSize(24);
            return;
        }
    }
```

程序在 Genymotion 中的运行效果如图 5.3 所示。

图 5.3 使用布局文件 activity_main.xml 注册事件监听器实例运行效果

第 6 章 Android 服务

Service（服务）是能够在后台长时间运行且不提供用户界面的应用程序组件。例如，Service 能在后台播放音乐、处理网络事务或执行文件输入/输出等操作。Service 是 Android 四大组件中与 Activity 最相似的组件，它们都代表可执行的程序。Service 与 Activity 的区别是，Service 一直在后台运行，没有用户界面，所以绝对不会到前台来。

6.1 Service 的分类

Service 按其状态可以分为两种类型，如表 6.1 所示。

表 6.1 Service 的分类

状 态	描 述
Started（启动）	当应用程序组件（如 Activity）通过调用 startService()方法启动 Service 时，Service 处于 Started 状态。一旦 Service 被启动，Service 就能在后台无限期地运行，即使启动它的组件已经被销毁。例如，Service 可以通过网络下载或上传文件。如果操作完成，Service 就需要停止运行
Bound（绑定）	当应用程序组件通过调用 bindService()方法绑定到 Service 时，Service 处于 Bound 状态。绑定 Service 可提供客户端/服务器接口，以允许组件与服务交互、发送请求、获得结果，甚至使用进程间通信（IPC）跨进程完成这些操作。仅当其他应用程序与 Service 绑定时，绑定 Service 才会运行。多个组件可以被一次性绑定到一个 Service 上，当它们都解绑时，Service 会被销毁

6.2 Service 的生命周期

Service 与 Activity 一样具有自己的生命周期。通过生命周期中的回调方法，用户可以监视 Service 的状态变化。根据应用程序启动 Service 的方式不同，Service 的生命周期也略有差异。

如果应用程序通过 startService()方法启动 Service，则 Service 的生命周期如图 6.1（左）所示。

如果应用程序通过 bindService()方法启动 Service，则 Service 的生命周期如图 6.1（右）所示。

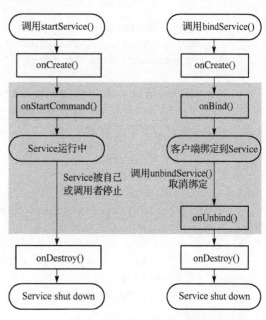

图 6.1 Service 的生命周期

6.3 Service 生命周期中的回调方法

为了创建 Service，需要创建一个 Service 类的子类。在实现类中，需要重写一些处理 Service 生命周期的回调方法，并根据需要提供组件绑定到服务的机制。需要重写的重要回调方法如表 6.2 所示。

表 6.2 Service 生命周期中的回调方法

回调方法	描述
onStartCommand()	当其他组件（如 Activity）调用 startService()方法启动该 Service 时，系统会回调该方法。如果任务完成且需要停止服务，则可以通过 stopSelf()或 stopService()方法实现
onBind()	当其他组件（如 Activity）调用 bindService()方法与 Service 绑定时，系统会调用该方法。在该方法的实现中，开发人员必须通过返回 IBinder 对象来提供其他组件与 Service 通信的接口。该方法必须实现，但是如果不想实现绑定，则需要返回 null
onUnbind()	当该 Service 上绑定的所有其他组件都断开连接时，将会回调该方法
onRebind()	当该 Service 和旧组件之间的所有绑定在 onUnbind()方法中全部结束之后，如果一个新的组件要绑定到该 Service，就会启动该方法
onCreate()	当该 Service 第一次被创建之后，将立即回调该方法。如果 Service 已经运行，则该方法不会被调用
onDestroy()	在该 Service 被关闭之前，将会回调该方法

下面的类定义了一个 Service 组件，重写了 Service 组件的 onCreate()、onStartCommand()、onBind()、onUnbind()、onRebind()和 onDestroy()方法，并且在重写这些方法时，只是简单地输出了一条状态信息的字符串：

```java
package com.example.helloservice;

import android.app.Service;
import android.content.Intent;
import android.os.IBinder;

public class HelloService extends Service {
    /** indicates how to behave if the service is killed */
    int mStartMode;

    /** interface for clients that bind */
    IBinder mBinder;

    /** indicates whether onRebind should be used */
    boolean mAllowRebind;

    /** Called when the service is being created. */
    @Override
    public void onCreate() {

    }

    /** The service is starting, due to a call to startService() */
    @Override
```

```java
public int onStartCommand(Intent intent, int flags, int startId) {
    return mStartMode;
}

/** A client is binding to the service with bindService() */
@Override
public IBinder onBind(Intent intent) {
    return mBinder;
}

/** Called when all clients have unbound with unbindService() */
@Override
public boolean onUnbind(Intent intent) {
    return mAllowRebind;
}

/** Called when a client is binding to the service with bindService()*/
@Override
public void onRebind(Intent intent) {

}

/** Called when The service is no longer used and is being destroyed */
@Override
public void onDestroy() {

}
}
```

【例 6.1】 Service 实例

这个例子将通过几个简单的步骤来展示如何创建一个 Android 服务，如表 6.3 所示。

表 6.3 Service 实例步骤

步骤	描述
1	打开 Android Studio，创建一个 Android 应用，选择 "Empty Activity" 选项，在 "Name" 文本框中输入项目名，在 "Package name" 文本框中输入包名
2	在 "Minimum SDK" 下拉列表中选择 "API 18: Android 4.3（Jelly Bean）" 选项或其他 SDK
3	在 "Language" 下拉列表中选择 "Java" 选项
4	选择保存路径，单击 "Finish" 按钮
5	右击工程项目的 com.example.helloservicedemo 目录，在弹出的快捷菜单中选择 "New" → "Service" → "Service" 命令，新建一个服务类并将其命名为 MyService.java。该文件将实现 Android 服务的相关方法
6	在工程项目中找到 ras/layout 目录中的 activity_main.xml 文件，并在其中添加两个 Button（Start Service 和 Stop Service）控件，按如图 6.2 所示的运行效果进行摆放
7	修改 java/com.example.helloservicedemo 目录下的 Java 文件 MainActivity.java 的代码，为其添加 startService()和 stopService()方法
8	使用<service.../>标签在 AndroidManifest.xml 文件中注册服务
9	启动 Genymotion 模拟器，然后在 Android 工程项目中进行如下代码修改

修改工程项目的 res/layout 目录下的布局文件 activity_main.xml，修改代码如下：

```xml
<?xml version="1.0" encoding="utf-8"?>
<LinearLayout xmlns:android="http://schemas.android.com/apk/res/android"
    android:layout_width="fill_parent"
    android:layout_height="fill_parent"
    android:orientation="vertical" >

    <Button android:id="@+id/btnStartService"
        android:layout_width="fill_parent"
        android:layout_height="wrap_content"
        android:text="Start Service"
        android:onClick="startService" />

    <Button android:id="@+id/btnStopService"
        android:layout_width="fill_parent"
        android:layout_height="wrap_content"
        android:text="Stop Service"
        android:onClick="stopService" />

</LinearLayout>
```

修改工程项目的 java/com.example.helloservicedemo 目录下的 Service 类 Java 文件 MyService.java，修改代码如下：

```java
package com.example.helloservicedemo;

import android.app.Service;
import android.content.Intent;
import android.os.IBinder;
import android.widget.Toast;

public class MyService extends Service {
    @Override
    public IBinder onBind(Intent arg0) {
        return null;
    }

    @Override
    public int onStartCommand(Intent intent, int flags, int startId) {
        // Let it continue running until it is stopped.
        Toast.makeText(this, "Service Started", Toast.LENGTH_LONG).show();
        return START_STICKY;
    }
    @Override
    public void onDestroy() {
        super.onDestroy();
        Toast.makeText(this, "Service Destroyed", Toast.LENGTH_LONG).show();
    }
}
```

修改工程项目的 java/com.example.helloservicedemo 目录下的 Java 文件 MainActivity.java，修改代码如下：

```java
package com.example.helloservicedemo;

import android.app.Activity;
import android.content.Intent;
import android.os.Bundle;
import android.view.View;

public class MainActivity extends Activity {

    @Override
    protected void onCreate(Bundle savedInstanceState) {
        super.onCreate(savedInstanceState);
        setContentView(R.layout.activity_main);
    }
    // Method to start the service
    public void startService(View view) {
        startService(new Intent(getBaseContext(), MyService.class));
    }

    // Method to stop the service
    public void stopService(View view) {
        stopService(new Intent(getBaseContext(), MyService.class));
    }
}
```

修改工程项目的 manifests 目录下的配置文件 AndroidManifest.xml，修改代码如下：

```xml
<?xml version="1.0" encoding="utf-8"?>
<manifest xmlns:android="http://schemas.android.com/apk/res/android"
    package="com.example.helloservicedemo" >

    <application
        android:allowBackup="true"
        android:icon="@mipmap/ic_launcher"
        android:label="@string/app_name"
        android:supportsRtl="true"
        android:theme="@style/AppTheme" >
        <activity android:name=".MainActivity" >
            <intent-filter>
                <action android:name="android.intent.action.MAIN" />

                <category android:name="android.intent.category.LAUNCHER" />
            </intent-filter>
        </activity>

        <service
            android:name=".MyService"
```

```
                android:enabled="true"
                android:exported="true" >
        </service>
    </application>

</manifest>
```

在 Genymotion 中运行程序,单击"Start Service"按钮,效果如图 6.2 所示。然后单击"Stop Service"按钮,效果如图 6.3 所示。

图 6.2 Service 实例运行效果(1)

图 6.3 Service 实例运行效果(2)

第 7 章　Android 广播接收器

Broadcast Receiver（广播接收器）也是 Android 系统四大组件之一，这种组件本质上是一个全局的监听器，用于监听系统全局的广播信息。这些信息就是程序（包括用户开发的程序和系统内建的程序）所发出的 Broadcast Intent（广播意图）。比如，当从网上下载完数据后，应用程序可以发出广播信息通知其他程序，数据已经下载完成并可以使用了。这时，Broadcast Receiver 可以监听到这个信息并采取相应的行动。

开发者可以通过以下两个步骤来创建 Broadcast Receiver，并使其响应与之相匹配的系统 Broadcast Intent。
- 创建 Broadcast Receiver。
- 注册 Broadcast Receiver。

也可以在以上两个步骤之间增加一个步骤，就是当执行用户自定义的 Broadcast Intent 时，需要创建这个自定义的 Broadcast Intent，然后将其与 Broadcast Receiver 一起注册。

7.1　创建 Broadcast Receiver

可以通过创建一个 Broadcast Receiver 子类的方法来创建一个 Broadcast Receiver。然后，重写它的 onReceive()方法，将 Intent 作为其参数来传递消息，例如：

```
public class MyReceiver extends BroadcastReceiver {
  @Override
  public void onReceive(Context context, Intent intent) {
    Toast.makeText(context, "Intent Detected.", Toast.LENGTH_LONG).show();
  }
}
```

7.2　注册 Broadcast Receiver

一旦实现了 Broadcast Receiver，接下来就应该指定该 Broadcast Receiver 所能匹配的 Intent，即注册 Broadcast Receiver，如图 7.1 所示。注册 Broadcast Receiver 需要考虑以下两种情况。
- 接收用户自定义 Broadcast Intent 消息。
- 接收系统广播消息。

图 7.1　注册 Broadcast Receiver

7.2.1 接收用户自定义 Broadcast Intent 消息

如果用户的应用程序组件（如 Activity）生成了一个自定义 Broadcast Intent 消息，则可以通过 sendBroadcast()方法将其发送出去，例如：

```
Intent intent = new Intent();
intent.setAction("com.example.CUSTOM_INTENT");
sendBroadcast(intent);
```

在 AndroidManifest.xml 文件中对与之相匹配的 Broadcast Receiver 进行注册，例如：

```xml
<application
  android:icon="@drawable/ic_launcher"
  android:label="@string/app_name"
  android:theme="@style/AppTheme" >
  <receiver android:name="MyReceiver">

    <intent-filter>
      <action android:name="com.example.CUSTOM_INTENT">
      </action>
    </intent-filter>

  </receiver>
</application>
```

在每次 Broadcast 事件发生后，系统就会创建对应的 Broadcast Receiver 实例，并自动触发它的 onReceive()方法。在 onReceive()方法执行完成后，Broadcast Receiver 实例就会被销毁。

如果 Broadcast Receiver 的 onReceive()方法不能在 10s 内执行完成，Android 就会认为该程序无响应。所以，不要在 Broadcast Receiver 的 onReceive()方法中执行一些耗时的操作，否则会弹出程序无响应对话框。如果确实需要使用 Broadcast Receiver 完成一项比较耗时的操作，则可以考虑通过 Intent 启动一个 Service 来完成该操作。

7.2.2 接收系统广播消息

除了接收用户发送的广播消息，Broadcast Receiver 还有一个重要的用途：接收系统广播消息。如果应用程序需要在系统特定时刻执行某些操作，就可以通过监听系统广播消息来实现。而若要注册监听系统广播消息的 Broadcast Receiver，则需要在 AndroidManifest.xml 文件中对与系统的广播 Action 常量匹配的 Broadcast Receiver 进行注册，例如：

```xml
<application
  android:icon="@drawable/ic_launcher"
  android:label="@string/app_name"
  android:theme="@style/AppTheme" >
  <receiver android:name="MyReceiver">

    <intent-filter>
      <action android:name="android.intent.action.BOOT_COMPLETED">
      </action>
    </intent-filter>
```

```
        </receiver>
    </application>
```

Android 的大量系统事件都会对外发送标准广播。Android 常见的广播 Action 常量如表 7.1 所示。

表 7.1 常见的广播 Action 常量

广播 Action 常量	描　　述
android.intent.action.BATTERY_CHANGED	当电池电量改变时的广播
android.intent.action.BATTERY_LOW	当电池电量低时的广播
android.intent.action.BATTERY_OKAY	当电池电量充足时的广播，即电池电量从低变化到饱满时会发出广播
android.intent.action.BOOT_COMPLETED	当完成系统启动后，这个动作会被广播一次（只有一次）
android.intent.action.SCREEN_ON	当屏幕被打开时的广播
android.intent.action.CALL	当用户根据指明的信息向某人拨打电话时的广播
android.intent.action.CALL_BUTTON	当用户单击"拨号"按钮时的广播
android.intent.action.DATE_CHANGED	当系统日期被改变时的广播
android.intent.action.REBOOT	当系统重启时的广播

注意：更多的 Action 常量请参考 Android 官方 API 文档中关于 Intent 的说明。

【**例 7.1**】 监听用户自定义 Broadcast Intent 消息实例

这个例子将通过几个简单的步骤来展示如何创建一个 Android Broadcast Receiver，用于监听用户自定义 Broadcast Intent 消息，如表 7.2 所示。

表 7.2 监听用户自定义 Broadcast Intent 消息实例步骤

步　骤	描　　述
1	打开 Android Studio，创建一个 Android 应用，选择 "Empty Activity" 选项，在 "Name" 文本框中输入项目名，在 "Package name" 文本框中输入包名
2	在 "Minimum SDK" 下拉列表中选择 "API 18: Android 4.3（Jelly Bean）"选项或其他 SDK
3	在 "Language" 下拉列表中选择 "Java" 选项
4	选择保存路径，单击 "Finish" 按钮
5	右击工程项目的 com.example.broadcastcustom 目录，在弹出的快捷菜单中选择 "New" → "Other" → "Broadcast Receiver" 命令，新建一个 Broadcast Receiver 类并将其命名为 MyReceiver.java。该文件将定义一个 Broadcast Receiver
6	将工程项目的 res/mipmap 目录下的图片文件 ic_launcher.png 复制、粘贴到 res/drawable 目录下
7	在工程项目中找到 ras/layout 目录中的 activity_main.xml 文件，并在其中添加一个 TextView、一个 ImageButton 和一个 Button（Broadcast Intent）控件，按如图 7.2 所示的运行效果进行摆放
8	修改 java/com.example.broadcastcustom 目录下的 Java 文件 MainActivity.java 的代码，为其添加自定义 Broadcast Intent 消息，并发送
9	使用<receiver.../>标签在 AndroidManifest.xml 文件中注册 Broadcast Receiver，并使用<intent-filter.../>标签设置 Broadcast Receiver 的过滤器
10	启动 Genymotion 模拟器，然后在 Android 工程项目中进行如下代码修改

修改工程项目的 java/com.example.broadcastcustom 目录下的服务类 Java 文件 MyReceiver.java，修改代码如下：

```
package com.example.broadcastcustom;
```

```java
import android.content.BroadcastReceiver;
import android.content.Context;
import android.content.Intent;
import android.widget.Toast;

public class MyReceiver extends BroadcastReceiver {
    public MyReceiver(){}
    @Override
    public void onReceive(Context context, Intent intent) {
        Toast.makeText(context,"Intent Detected.",Toast.LENGTH_LONG).show();
    }
}
```

修改新建工程项目的 res/layout 目录下的布局文件 activity_main.xml，修改代码如下：

```xml
<?xml version="1.0" encoding="utf-8"?>
<RelativeLayout xmlns:android="http://schemas.android.com/apk/res/android"
    xmlns:tools="http://schemas.android.com/tools"
    android:layout_width="match_parent"
    android:layout_height="match_parent"
    android:paddingLeft="@dimen/activity_horizontal_margin"
    android:paddingRight="@dimen/activity_horizontal_margin"
    android:paddingTop="@dimen/activity_vertical_margin"
    android:paddingBottom="@dimen/activity_vertical_margin"
    tools:context=".MainActivity">

    <TextView
        android:id="@+id/textView1"
        android:layout_width="wrap_content"
        android:layout_height="wrap_content"
        android:text="Example of Broadcast"
        android:layout_alignParentTop="true"
        android:layout_centerHorizontal="true"
        android:textSize="30dp" />

    <ImageButton
        android:layout_width="wrap_content"
        android:layout_height="wrap_content"
        android:id="@+id/imageButton"
        android:src="@drawable/ic_launcher"
        android:layout_centerVertical="true"
        android:layout_centerHorizontal="true" />

    <Button
        android:layout_width="wrap_content"
        android:layout_height="wrap_content"
        android:id="@+id/button"
        android:text="Broadcast Intent"
        android:onClick="broadcastIntent"
```

```xml
            android:layout_below="@+id/imageButton"
            android:layout_alignEnd="@+id/imageButton"
            android:layout_alignStart="@+id/imageButton" />
</RelativeLayout>
```

修改工程项目的 java/com.example.broadcastcustom 目录下的 Java 文件 MainActivity.java，修改代码如下：

```java
package com.example.broadcastcustom;

import android.app.Activity;
import android.content.Intent;
import android.os.Bundle;
import android.view.View;
import android.widget.Button;

public class MainActivity extends Activity {

    @Override
    protected void onCreate(Bundle savedInstanceState) {
        super.onCreate(savedInstanceState);
        setContentView(R.layout.activity_main);
        Button send= (Button)findViewById(R.id.button);
        send.setOnClickListener(new View.OnClickListener() {
            @Override
            public void onClick(View view) {
                Intent intent = new Intent();
                intent.setAction("com.example.action.DEMO_BROADCAST");
                sendBroadcast(intent);
            }
        });
    }
}
```

修改工程项目的 manifests 目录下的配置文件 AndroidManifest.xml，修改代码如下：

```xml
<?xml version="1.0" encoding="utf-8"?>
<manifest xmlns:android="http://schemas.android.com/apk/res/android"
    package="com.example.broadcastcustom" >

    <application
        android:allowBackup="true"
        android:icon="@mipmap/ic_launcher"
        android:label="@string/app_name"
        android:supportsRtl="true"
        android:theme="@style/AppTheme" >
        <activity android:name=".MainActivity" >
            <intent-filter>
                <action android:name="android.intent.action.MAIN" />
```

```xml
        <category android:name="android.intent.category.LAUNCHER" />
      </intent-filter>
    </activity>

    <receiver android:name=".MyReceiver" >
      <intent-filter android:priority="20" >
        <action android:name="com.example.action.DEMO_BROADCAST" >
        </action>
      </intent-filter>
    </receiver>

  </application>

</manifest>
```

在 Genymotion 中运行程序，单击"Broadcast Intent"按钮，效果如图 7.2 所示。

图 7.2　监听用户自定义 Broadcast Intent 消息实例运行效果

【例 7.2】　监听系统广播消息实例

这个例子将通过几个简单的步骤来展示如何创建一个 Android Broadcast Receiver，用于监听手机电池状态的系统广播消息，如表 7.3 所示。

表 7.3　监听系统广播消息实例步骤

步　骤	描　述
1	打开 Android Studio，创建一个 Android 应用，选择"Empty Activity"选项，在"Name"文本框中输入项目名，在"Package name"文本框中输入包名
2	在"Minimum SDK"下拉列表中选择"API 18: Android 4.3（Jelly Bean）"选项或其他 SDK

步骤	描述
3	在"Language"下拉列表中选择"Java"选项
4	选择保存路径,单击"Finish"按钮
5	右击工程项目的 com.example.broadcastcustom 目录,在弹出的快捷菜单中选择"New"→"Other"→"Broadcast Receiver"命令,新建一个 Broadcast Receiver 类并将其命名为 MyReceiver.java。该文件将定义一个 Broadcast Receiver
6	使用<receiver.../>标签在 AndroidManifest.xml 文件中注册 Broadcast Receiver,并使用<intent-filter.../>标签设置 Broadcast Receiver 的过滤器
7	启动 Genymotion 模拟器,然后在 Android 工程项目中进行如下代码修改

修改工程项目的 java/com.example.broadcastsystem 目录下的服务类 Java 文件 **MyReceiver.java**,修改代码如下:

```java
package com.example.broadcastsystem;

import android.content.BroadcastReceiver;
import android.content.Context;
import android.content.Intent;
import android.os.Bundle;
import android.widget.Toast;

public class MyReceiver extends BroadcastReceiver {
    @Override
    public void onReceive(Context context, Intent intent) {
        if (Intent.ACTION_BATTERY_OKAY.equals(intent.getAction())) {
            Toast.makeText(context,"电量已恢复,可以使用!",Toast.LENGTH_LONG).show();
        }
        if (Intent.ACTION_BATTERY_LOW.equals(intent.getAction())) {
            Toast.makeText(context,"电量过低,请尽快充电!",Toast.LENGTH_LONG).show();
        }
        if (Intent.ACTION_BATTERY_CHANGED.equals(intent.getAction())) {
            Bundle bundle = intent.getExtras();
            // 获取当前电量
            int current = bundle.getInt("level");
            // 获取总电量
            int total = bundle.getInt("scale");
            StringBuffer sb = new StringBuffer();
            sb.append("当前电量为: " + current * 100 / total + "%" + " ");
            // 如果当前电量小于总电量的15%
            if (current * 1.0 / total < 0.15) {
                sb.append("电量过低,请尽快充电! ");
            } else {
                sb.append("电量足够,请放心使用! ");
            }
            Toast.makeText(context, sb.toString(), Toast.LENGTH_LONG).show();
        }
    }
}
```

修改工程项目的 manifests 目录下的配置文件 AndroidManifest.xml，修改代码如下：

```xml
<?xml version="1.0" encoding="utf-8"?>
<manifest xmlns:android="http://schemas.android.com/apk/res/android"
    package="com.example.broadcastsystem" >
    <uses-permission android:name="android.permission.BATTERY_STATS"/>
    <application
        android:allowBackup="true"
        android:icon="@mipmap/ic_launcher"
        android:label="@string/app_name"
        android:supportsRtl="true"
        android:theme="@style/AppTheme" >

        <activity android:name=".MainActivity" >
            <intent-filter>
                <action android:name="android.intent.action.MAIN" />

                <category android:name="android.intent.category.LAUNCHER" />
            </intent-filter>
        </activity>

        <receiver android:name=".MyReceiver" >
            <intent-filter>
                <action android:name="android.intent.action.BATTERY_CHANGED" />
                <action android:name="android.intent.action.BATTERY_OKAY"/>
                <action android:name="android.intent.action.BATTERY_LOW"/>
            </intent-filter>
        </receiver>
    </application>

</manifest>
```

在 Genymotion 模拟器中单击电池图标🔋，可弹出如图 7.3 所示的电池调节界面。

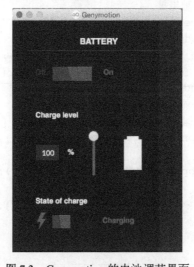

图 7.3　Genymotion 的电池调节界面

如图 7.3 所示，调节 Charge level（电池量）的百分比，当电量低于 15%时，将会弹出"电量过低，请尽快充电！"的提示信息，如图 7.4 所示。

图 7.4　监听系统广播消息实例运行效果

这里只演示了电量低的情况，读者可以调节 Charge level 百分比以测试其他几种电池状态的运行提示。

注意：该程序在模拟器中运行时，可能会有延时的情况出现。请多尝试几次或者在真机中运行该程序。

第 8 章 ContentProvider 实现数据共享

当在系统中部署越来越多的 Android 应用后,有时需要在不同的应用之间共享数据。比如,现在有一个短信接收应用,用户想把发送短信的人添加到联系人管理应用中,就需要在不同的应用之间共享数据。对于这种需求,相应的操作方法有很多种,比如,使用 SharedPreferences、文件或数据库等直接操作另一个应用程序所记录的数据。但这种方式显得太杂乱,因为不同的应用程序记录数据的方式差别太大,不利于在应用程序之间进行数据交换。

8.1 ContentProvider 概述

为了在应用程序之间进行数据交换,Android 提供了 ContentProvider。ContentProvider 是在不同应用程序之间进行数据交换的标准 API,当一个应用程序需要把自己的数据交给其他应用程序使用时,该应用程序就可以通过 ContentProvider 来实现;其他应用程序可以通过 ContentResolver 来操作 ContentProvider 提供的数据。ContentProvider 也可以把数据保存在数据库、文件甚至网络中。ContentProvider 运行模式如图 8.1 所示。

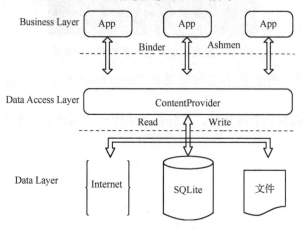

图 8.1 ContentProvider 运行模式

一旦某个应用程序通过 ContentProvider 共享了自己的数据操作接口,那么不管该应用程序是否启动,其他应用程序都可以通过该接口来操作共享的内部数据。ContentProvider 提供了用于增加数据的 insert()、用于修改数据的 update()、用于删除数据的 delete()和用于查询数据的 query()等数据操作方法。

要创建一个自定义的 ContentProvider 类,可以通过继承 Android 提供的 ContentProvider 基类来实现,例如:

```
public class My Application extends  ContentProvider {

}
```

ContentProvider 也是 Android 应用的四大组件之一,它与 Activity、Service 和 BroadcastReceiver

相似，都需要在 AndroidManifest.xml 文件中进行注册，例如：

```xml
<provider
    android:name=".StudentsProvider"
    android:authorities="com.example.provider.college"
    android:enabled="true"
    android:exported="true" >
</provider>
```

从上面的注册代码可以看出，在注册 ContentProvider 时通常指定如表 8.1 所示的属性。

表 8.1 在注册 ContentProvider 时通常指定的属性

属 性	描 述
name	指定该 ContentProvider 的实现类的类名
authorities	指定该 ContentProvider 对应的 URI
enabled	指定该 ContentProvider 是否可用
exported	指定该 ContentProvider 是否允许其他应用程序调用

8.2 URI 简介

每个 ContentProvider 都需要提供公共的 URI 来唯一标识其数据集。管理多个数据集的 ContentProvider 为每个数据集提供了单独的 URI。URI 的基本格式如下：

<prefix>://<authority>/<data_type>/<id>

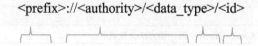

如 content:// com.example.provider.college /people/2

表 8.2 详细介绍了 URI 的每个组成部分。

表 8.2 URI 的组成部分详解

组 成 部 分	描 述
<prefix>	标准的前缀，始终设置为 content://
<authority>	指定 ContentProvider 的名称，如 contacts 或 browser 等（系统 ContentProvider）。对于第三方应用，该部分应该是完整的 ContentProvider 类名，如 com.example.provider.college
<data_type>	指定 ContentProvider 的被请求数据的路径。例如，用户需要从 contacts 这个系统 ContentProvider 中查询所有联系人，则 URI 为 content://contacts/people
<id>	被请求的特定记录的 id 值。例如，用户需要从 contacts 这个系统 ContentProvider 中查询第 5 个联系人，则 URI 为 content://contacts/people/5

8.3 创建 ContentProvider

创建 ContentProvider 只需要进行如下两个步骤。
- 像前文提到的一样，要创建一个自定义 ContentProvider 类，可以通过继承 Android 提供的 ContentProvider 基类来实现。
- 在 AndroidManifest.xml 文件中注册该 ContentProvider，指定 android:authorities 属性，也就是为 ContentProvider 绑定一个 URI。

在上面两个步骤中，自定义 ContentProvider 类实现的 insert()、update()、delete()和 query()方法并不是供该应用本身调用的，而是供其他应用调用的。当其他应用通过 ContentResolver 调用 insert()、update()、delete()和 query()方法执行数据访问时，实际上就是调用指定 URI 对应的 ContentProvider 的 insert()、update()、delete()和 query()方法。

如何重写 ContentProvider 的 insert()、update()、delete()和 query()等方法，完全取决于程序员。下面通过图 8.2 和表 8.3 对 ContentProvider 类中的重写方法进行简单描述。

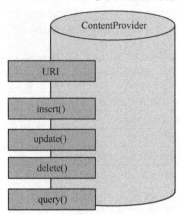

图 8.2　ContentProvider 类的结构

表 8.3　ContentProvider 子类中的重写方法

重写方法	描述
onCreate()	当开启 ContentProvider 时，该方法会被调用
query()	该方法用于接收客户端查询请求，并返回一个 Cursor 对象作为结果
insert()	该方法用于为 ContentProvider 添加一条新记录
delete()	该方法用于从 ContentProvider 中删除一条记录
update()	该方法用于修改一条 ContentProvider 记录
getType()	该方法用于返回给定 URI 指向数据的 MIME 类型

【例 8.1】　创建 ContentProvider 实例

这个例子将通过几个简单的步骤来展示如何创建一个 ContentProvider，如表 8.4 所示。

表 8.4　创建 ContentProvider 实例步骤

步骤	描述
1	打开 Android Studio，创建一个 Android 应用，选择"Empty Activity"选项，在"Name"文本框中输入项目名，在"Package name"文本框中输入包名
2	在"Minimum SDK"下拉列表中选择"API 18: Android 4.3（Jelly Bean）"选项或其他 SDK
3	在"Language"下拉列表中选择"Java"选项
4	选择保存路径，单击"Finish"按钮
5	右击工程项目的 com.example.contentprovider 目录，在弹出的快捷菜单中选择"New"→"Other"→"Content Provider"命令，新建一个服务类并将其命名为 StudentsProvider.java。该文件将定义 ContentProvider 和相关方法
6	将工程项目的 res/mipmap 目录下的图片文件 ic_launcher.png 复制、粘贴到 res/drawable 目录下
7	在工程项目中找到 ras/layout 目录中的 activity_main.xml 文件，并在其中添加一个 TextView、一个 ImageButton、两个 EditText（Name 和 Grade）和两个 Button（Add Name 和 Retrive student）控件，按如图 8.3 所示的运行效果进行摆放

续表

步骤	描述
8	修改 java/com.example.contentprovider 目录下的 Java 文件 MainActivity.java 的代码，为其添加 onClickAddName() 和 onClickRetrieveStudents()方法
9	使用<provider…/>标签在 AndroidManifest.xml 文件中注册 ContentProvider
10	启动 Genymotion 模拟器，然后在 Android 工程项目中进行如下代码修改

修改工程项目的 res/layout 目录下的布局文件 activity_main.xml，修改代码如下：

```xml
<?xml version="1.0" encoding="utf-8"?>
<RelativeLayout xmlns:android="http://schemas.android.com/apk/res/android"
    xmlns:tools="http://schemas.android.com/tools"
    android:layout_width="match_parent"
    android:layout_height="match_parent"
    android:paddingLeft="@dimen/activity_horizontal_margin"
    android:paddingRight="@dimen/activity_horizontal_margin"
    android:paddingTop="@dimen/activity_vertical_margin"
    android:paddingBottom="@dimen/activity_vertical_margin"
    tools:context=".MainActivity">

    <TextView
        android:id="@+id/textView1"
        android:layout_width="wrap_content"
        android:layout_height="wrap_content"
        android:text="Content provider"
        android:layout_alignParentTop="true"
        android:layout_centerHorizontal="true"
        android:textSize="30dp" />

    <ImageButton
        android:layout_width="wrap_content"
        android:layout_height="wrap_content"
        android:id="@+id/imageButton"
        android:src="@drawable/ic_launcher"
        android:layout_centerHorizontal="true"
        android:layout_below="@+id/textView1" />

    <Button
        android:layout_width="wrap_content"
        android:layout_height="wrap_content"
        android:id="@+id/button2"
        android:text="Add Name"
        android:layout_below="@+id/editText3"
        android:onClick="onClickAddName"
        android:layout_alignEnd="@+id/editText3"
        android:layout_alignStart="@+id/editText3" />

    <EditText
```

```xml
        android:layout_width="wrap_content"
        android:layout_height="wrap_content"
        android:id="@+id/editText2"
        android:hint="Name"
        android:textColorHint="@android:color/holo_blue_light"
        android:layout_below="@+id/imageButton"
        android:layout_alignEnd="@+id/editText3"
        android:layout_alignStart="@+id/editText3" />

    <EditText
        android:layout_width="wrap_content"
        android:layout_height="wrap_content"
        android:id="@+id/editText3"
        android:hint="Grade"
        android:textColorHint="@android:color/holo_blue_bright"
        android:layout_below="@+id/editText2"
        android:layout_alignEnd="@+id/imageButton"
        android:layout_alignStart="@+id/imageButton" />

    <Button
        android:layout_width="wrap_content"
        android:layout_height="wrap_content"
        android:text="Retrive student"
        android:id="@+id/button"
        android:layout_below="@+id/button2"
        android:layout_alignRight="@+id/editText3"
        android:layout_alignEnd="@+id/editText3"
        android:layout_alignLeft="@+id/button2"
        android:layout_alignStart="@+id/button2"
        android:onClick="onClickRetrieveStudents"/>
</RelativeLayout>
```

修改工程项目的 java/com.example.contentprovider 目录下的服务类 Java 文件 StudentsProvider.java，修改代码如下：

```java
package com.example.contentprovider;

import android.content.ContentProvider;
import android.content.ContentUris;
import android.content.ContentValues;
import android.content.Context;
import android.content.UriMatcher;
import android.database.Cursor;
import android.database.SQLException;
import android.database.sqlite.SQLiteDatabase;
import android.database.sqlite.SQLiteOpenHelper;
import android.database.sqlite.SQLiteQueryBuilder;
import android.net.Uri;
import android.text.TextUtils;
```

```java
import java.util.HashMap;

public class StudentsProvider extends ContentProvider {
    static final String PROVIDER_NAME = "com.example.provider.College";
    static final String URL = "content://" + PROVIDER_NAME + "/students";
    static final Uri CONTENT_URI = Uri.parse(URL);

    static final String _ID = "_id";
    static final String NAME = "name";
    static final String GRADE = "grade";

    private static HashMap<String, String> STUDENTS_PROJECTION_MAP;

    static final int STUDENTS = 1;
    static final int STUDENT_ID = 2;

    static final UriMatcher uriMatcher;
    static{
        uriMatcher = new UriMatcher(UriMatcher.NO_MATCH);
        uriMatcher.addURI(PROVIDER_NAME, "students", STUDENTS);
        uriMatcher.addURI(PROVIDER_NAME, "students/#", STUDENT_ID);
    }

    /**
     * Database specific constant declarations
     */
    private SQLiteDatabase db;
    static final String DATABASE_NAME = "College";
    static final String STUDENTS_TABLE_NAME = "students";
    static final int DATABASE_VERSION = 1;
    static final String CREATE_DB_TABLE =
        " CREATE TABLE " + STUDENTS_TABLE_NAME +
        " (_id INTEGER PRIMARY KEY AUTOINCREMENT, " +
        " name TEXT NOT NULL, " +
        " grade TEXT NOT NULL);";

    /**
     * Helper class that actually creates and manages
     * the provider's underlying data repository.
     */
    private static class DatabaseHelper extends SQLiteOpenHelper {
        DatabaseHelper(Context context){
            super(context, DATABASE_NAME, null, DATABASE_VERSION);
        }

        @Override
        public void onCreate(SQLiteDatabase db)
        {
            db.execSQL(CREATE_DB_TABLE);
```

```java
        }

        @Override
        public void onUpgrade(SQLiteDatabase db, int oldVersion, int newVersion) {
            db.execSQL("DROP TABLE IF EXISTS " + STUDENTS_TABLE_NAME);
            onCreate(db);
        }
    }

    @Override
    public boolean onCreate() {
        Context context = getContext();
        DatabaseHelper dbHelper = new DatabaseHelper(context);

        /**
         * Create a write able database which will trigger its
         * creation if it doesn't already exist.
         */
        db = dbHelper.getWritableDatabase();
        return (db == null)? false:true;
    }

    @Override
    public Uri insert(Uri uri, ContentValues values) {
        /**
         * Add a new student record
         */
        long rowID = db.insert(    STUDENTS_TABLE_NAME, "", values);

        /**
         * If record is added successfully
         */

        if (rowID > 0)
        {
           Uri _uri = ContentUris.withAppendedId(CONTENT_URI, rowID);
           getContext().getContentResolver().notifyChange(_uri, null);
           return _uri;
        }
        throw new SQLException("Failed to add a record into " + uri);
    }

    @Override
    public Cursor query(Uri uri, String[] projection, String selection,
        String[] selectionArgs, String sortOrder) {
        SQLiteQueryBuilder qb = new SQLiteQueryBuilder();
        qb.setTables(STUDENTS_TABLE_NAME);

        switch (uriMatcher.match(uri)) {
```

```java
        case STUDENTS:
            qb.setProjectionMap(STUDENTS_PROJECTION_MAP);
            break;

        case STUDENT_ID:
            qb.appendWhere( _ID + "=" + uri.getPathSegments().get(1));
            break;

        default:
            throw new IllegalArgumentException("Unknown URI " + uri);
    }

    if (sortOrder == null || sortOrder == ""){
        /**
         * By default sort on student names
         */
        sortOrder = NAME;
    }
    Cursor c = qb.query(db,    projection,    selection, selectionArgs,
                null, null, sortOrder);

    /**
     * register to watch a content URI for changes
     */
    c.setNotificationUri(getContext().getContentResolver(), uri);
    return c;
}

@Override
public int delete(Uri uri, String selection, String[] selectionArgs) {
    int count = 0;

    switch (uriMatcher.match(uri)){
        case STUDENTS:
            count=db.delete(STUDENTS_TABLE_NAME, selection, selectionArgs);
            break;

        case STUDENT_ID:
            String id = uri.getPathSegments().get(1);
            count = db.delete( STUDENTS_TABLE_NAME, _ID +  " = " + id +
                    (!TextUtils.isEmpty(selection) ? " AND (" + selection
                    + ')' : ""), selectionArgs);
            break;

        default:
            throw new IllegalArgumentException("Unknown URI " + uri);
    }

    getContext().getContentResolver().notifyChange(uri, null);
```

```java
            return count;
    }

    @Override
    public int update(Uri uri, ContentValues values, String selection,
                    String[] selectionArgs) {
        int count = 0;

        switch (uriMatcher.match(uri)){
            case STUDENTS:
                count = db.update(STUDENTS_TABLE_NAME, values, selection,
                                selectionArgs);
                break;

            case STUDENT_ID:
                count = db.update(STUDENTS_TABLE_NAME, values, _ID + " = "
                        + uri.getPathSegments().get(1) +
                        (!TextUtils.isEmpty(selection) ? " AND (" +selection
                        + ')' : ""), selectionArgs);
                break;

            default:
                throw new IllegalArgumentException("Unknown URI " + uri );
        }
        getContext().getContentResolver().notifyChange(uri, null);
        return count;
    }

    @Override
    public String getType(Uri uri) {
        switch (uriMatcher.match(uri)){
            /**
             * Get all student records
             */
            case STUDENTS:
                return "vnd.android.cursor.dir/vnd.example.students";

            /**
             * Get a particular student
             */
            case STUDENT_ID:
                return "vnd.android.cursor.item/vnd.example.students";

            default:
                throw new IllegalArgumentException("Unsupported URI: "+uri);
        }
    }
}
```

第 8 章 ContentProvider 实现数据共享

修改工程项目的 java/com.example.contentprovider 目录下的 Java 文件 MainActivity.java，修改代码如下：

```java
package com.example.contentprovider;

import android.app.Activity;
import android.content.ContentValues;
import android.database.Cursor;
import android.net.Uri;
import android.os.Bundle;
import android.view.View;
import android.widget.EditText;
import android.widget.Toast;

public class MainActivity extends Activity {

    @Override
    protected void onCreate(Bundle savedInstanceState) {
        super.onCreate(savedInstanceState);
        setContentView(R.layout.activity_main);
    }
    public void onClickAddName(View view) {
        // Add a new student record
        ContentValues values = new ContentValues();

        values.put(StudentsProvider.NAME,
                ((EditText)findViewById(R.id.editText2)).getText().toString());

        values.put(StudentsProvider.GRADE,
                ((EditText)findViewById(R.id.editText3)).getText().toString());

        Uri uri = getContentResolver().insert(
                StudentsProvider.CONTENT_URI, values);

        Toast.makeText(getBaseContext(),
                uri.toString(), Toast.LENGTH_LONG).show();
    }

    public void onClickRetrieveStudents(View view) {

        // Retrieve student records
        String URL = "content://com.example.provider.College/students";

        Uri students = Uri.parse(URL);
        Cursor c = managedQuery(students, null, null, null, "name");

        if (c.moveToFirst()) {
            do{
```

```
                Toast.makeText(this,
                    c.getString(c.getColumnIndex(StudentsProvider._ID)) +
                        ", " + c.getString(c.getColumnIndex
                        (StudentsProvider.NAME)) +
                        ", " + c.getString(c.getColumnIndex
                        (StudentsProvider.GRADE)),
                    Toast.LENGTH_SHORT).show();
            } while (c.moveToNext());
        }
    }
}
```

修改工程项目的 manifests 目录下的配置文件 AndroidManifest.xml，修改代码如下：

```
<?xml version="1.0" encoding="utf-8"?>
<manifest xmlns:android="http://schemas.android.com/apk/res/android"
    package="com.example.contentprovider" >

    <application
        android:allowBackup="true"
        android:icon="@mipmap/ic_launcher"
        android:label="@string/app_name"
        android:supportsRtl="true"
        android:theme="@style/AppTheme" >
        <activity android:name=".MainActivity" >
            <intent-filter>
                <action android:name="android.intent.action.MAIN" />

                <category android:name="android.intent.category.LAUNCHER" />
            </intent-filter>
        </activity>

        <provider
            android:name=".StudentsProvider"
            android:authorities="com.example.provider.College"
            android:enabled="true"
            android:exported="true" >
        </provider>
    </application>

</manifest>
```

在 Genymotion 中运行程序，然后在"Name"编辑框中输入名字，在"Grade"编辑框中输入年级，单击"Add Name"按钮，效果如图 8.3 所示。

使用同样的方式连续输入多个学生信息，将会通过 insert()方法把学生信息保存到 ContentProvider 的数据库中，然后单击"Retrive student"按钮。数据库中保存的信息将会通过 query()方法返回并依次显示在屏幕中，运行效果如图 8.4 所示。

第 8 章 ContentProvider 实现数据共享

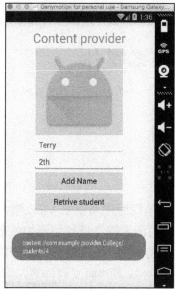

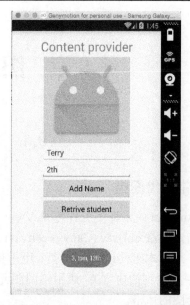

图 8.3　创建 ContentProvider 实例运行效果（1）　　图 8.4　创建 ContentProvider 实例运行效果（2）

第 9 章 图形、图片与多媒体

9.1 基础绘图

9.1.1 常用绘图类

在 Android 中，许多组件都是 View 类的子类，所以在绘图时应当继承 View 类，并重写其中的指定方法 onDraw(Canvas canvas)。如果想展示特定的图形或者设置图形的特效，则需要具备 4 个基本组件：Bitmap 类，用于保存像素；Canvas 类，用于保存绘图的回调结果（在 Bitmap 中要绘制的内容）；通用绘图图元，如 Rect、Path、text、Bitmap；Print 类，用于设置颜色及绘图的风格。

Android 中的常用绘图类包括以下几个。

（1）android.graphics.Bitmap：代表一张位图，所包含的图片可以由程序创建或者来自文件。

（2）android.graphics.Canvas：代表画布，可以通过设置画布的属性，如颜色或尺寸等，画出想画的东西。

（3）android.graphics.Paint：代表画笔，用于描述绘制图形的风格，可设置颜色、线宽、透明度等属性。在使用该类时，可以通过该类提供的构造方法创建对象。

在一般的绘图操作中，Canvas 类提供了一些常用的绘图方法，如绘制圆形、正方形、长方形、椭圆形等方法。使用这些方法可以直接画出图形。Canvas 类的常用绘图方法如表 9.1 所示。

表 9.1 Canvas 类的常用绘图方法

返回类型	方 法	简 述
boolean	clipRegion(Region region)	裁剪指定区域
	clipRect(int left, int top, int right, int bottom)	裁剪一个矩形区域
void	drawPoint(float x, float y, Paint paint)	绘制一个点
	drawPoints(float[] pts, int offset, int count, Paint paint)	绘制多个点
	drawLine(float startX, float startY, float stopX, float stopY, Paint paint)	绘制一条线
	drawLines(float[] pts, int offset, int count, Paint paint)	绘制多条线
	drawText(String text, float x, float y, Paint paint)	绘制字符串
	drawRect(float left, float top, float right, float bottom, Paint paint)	绘制矩形
	drawRoundRect(RectF rect, float rx, float ry, Paint paint)	绘制圆角矩形
	drawOval(float left, float top, float right, float bottom, Paint paint)	绘制椭圆
	drawArc(RectF oval, float startAngle, float sweepAngle, boolean useCenter, Paint paint)	绘制弧
	drawPaint(Paint paint)	在画布上绘制位图
	drawPath(Path path, Paint paint)	沿着指定 Path 绘制任意形状
	drawBitmap(Bitmap bitmap, float left, float top, Paint paint)	在指定坐标处绘制位图
	drawBitmap(Bitmap bitmap, Rect src, Rect dst, Paint paint)	在指定坐标处绘制从源位图中"挖取"的一块

第 9 章 图形、图片与多媒体

续表

返回类型	方 法	简 述
void	drawCircle(float cx, float cy, float radius, Paint paint)	在指定坐标处绘制圆形
	drawColor(int color)	用指定的颜色填充位图
	drawRGB(int r, int g, int b)	用指定的颜色 RGB 填充位图
	drawARGB(int a, int r, int g, int b)	用指定的颜色 ARGB 填充位图

在使用 Canvas 类进行绘图时，还需要结合 Paint 类。Paint 类代表画布上的画笔，用于绘图属性的设置。Paint 类的常用绘图方法如表 9.2 所示。

表 9.2　Paint 类的常用绘图方法

返回类型	方 法	简 述
boolean	setARGB(int a,int r,int g,int b)	设置颜色，参数值分别表示透明度、红色、绿色和蓝色值
	setAlpha(int a)	设置透明度
	setColor(int color)	设置颜色
	setAntiAlias(boolean aa)	设置是否使用抗锯齿功能
	setDither(boolean dither)	设置是否使用图片抖动处理
	setPathEffect(PathEffect effect)	设置绘制路径时的路径效果
	setShader(Shader shader)	设置渐变
	setShadowLayer(float radius,float dx,float dy,int color)	设置阴影
	setStyle(Paint.Style style)	设置画笔的样式风格
	setStrokeCap(Paint.Cap cap)	设置笔刷的图形样式
	setStrokeWidth(float width)	设置笔刷的宽度
	setSrokeJoin(Paint.Join join)	设置绘制时各图形的连接方式
	setXfermode(Xfermode xfermode)	设置图形重叠时的处理方式
	setTextAlign(Paint.Align align)	设置绘制文字的对齐方式
	setTextSize(float textSize)	设置绘制文字的字号大小
	setUnderlineText(boolean underlineText)	设置带有下画线的文字效果
	setStrikeThruText(boolean strikeThruText)	设置带有删除线的文字效果

9.1.2　绘制 2D 图形

在重写 onDraw()方法时，可以通过调用一些方法绘制基础的 2D 图形。下面详细介绍在 Android 中绘制简单的 2D 图形时用到的方法。

1．点

drawPoint(float x,float y,Paint paint)：在指定坐标处绘制一个点。

2．直线

drawLine(float startx,float starty,loat stopx,float stopy,Paint paint)：在坐标(startx,starty)和坐标(stopx,stopy)之间绘制一条直线。

3．矩形

drawRect(float x1,float y1,float x2,float y2,Paint paint)：绘制左上角顶点坐标为(x1,y1)，右下角顶点坐标为(x2,y2)的矩形。

4．多边形

drawVertices(VertexMode mode,int count,float[] pts,0,null,0,null,0,null,0,0,Paint paint)：绘制一

个多边形。其中，count 为坐标的个数，等于 pts 的大小。

5. 圆

drawCircle(float cx,float cy,float r,Paint paint)：绘制一个以坐标(cx,cy)为圆心、r 为半径的圆。

6. 弧

drawArc(Rect rect,float startAngle,float sweepAngle,bool useCenter,Paint paint)：在矩形 rect 内部（并不会绘制此矩形）绘制一个起始角度为 startAngle、结束角度为 sweepAngle 的弧。useCenter 用于决定这段弧是否会显示起点与终点的连线。

7. 文字

drawText(String s, float x, float y, Paint paint)：在坐标(x,y)处绘制字符串 s。

【例 9.1】 Canvas 类与 Paint 类实例

下面使用 Canvas 类与 Paint 类中的一些常用方法绘制一个 Android 机器人图标，具体实现方法如下：

```java
// 定义 MyView 类，此类继承 View 类
package com.example.canvaspaint;
import android.content.Context;
import android.graphics.Canvas;
import android.graphics.Color;
import android.graphics.Paint;
import android.graphics.Rect;
import android.graphics.RectF;
import android.util.AttributeSet;
import android.view.View;
public class MyView extends View {
    public MyView(Context context,AttributeSet attrs)    // 继承 View 类
    {
        super(context,attrs);
    }
    @Override
    protected void onDraw(Canvas canvas) {
        Paint paint=new Paint();                        // 采用默认设置创建一个画笔
        paint.setAntiAlias(true);                       // 使用抗锯齿功能
        paint.setColor(0xFF78C257);                     // 设置画笔的颜色为绿色
        // 绘制机器人的头
        RectF rectf_head = new RectF(110, 30, 200, 120);
        canvas.drawArc(rectf_head, -10, -160, false, paint);// 绘制弧
        // 绘制眼睛
        paint.setColor(Color.WHITE);                    // 设置画笔的颜色为白色
        canvas.drawCircle(135, 53, 4, paint);           // 绘制圆
        canvas.drawCircle(175, 53, 4, paint);           // 绘制圆
        // 绘制天线
        paint.setColor(0xFF78C257);                     // 设置画笔的颜色为绿色
        paint.setStrokeWidth(8);                        // 设置笔触的宽度
        canvas.drawLine(125, 20, 135, 35, paint);       // 绘制线
        canvas.drawLine(185, 20, 175, 35, paint);       // 绘制线
        // 绘制身体
        canvas.drawRect(110, 75, 200, 150, paint);      // 绘制矩形
```

```
            RectF rectf_body = new RectF(110,140,200,160);
            canvas.drawRoundRect(rectf_body, 10, 10, paint); // 绘制圆角矩形
            // 绘制胳膊
            RectF rectf_arm = new RectF(85,75,105,140);
            canvas.drawRoundRect(rectf_arm, 10, 10, paint);  // 绘制左侧的胳膊
            rectf_arm.offset(120, 0);                        //设置在 X 轴上偏移 120 像素
            canvas.drawRoundRect(rectf_arm, 10, 10, paint);  // 绘制右侧的胳膊
            // 绘制腿
            RectF rectf_leg = new RectF(125,150,145,200);
            canvas.drawRoundRect(rectf_leg, 10, 10, paint);  // 绘制左侧的腿
            rectf_leg.offset(40, 0);                         // 设置在 X 轴上偏移 40 像素
            canvas.drawRoundRect(rectf_leg, 10, 10, paint);  // 绘制右侧的腿
            // 写文字
            paint.setTextSize(50);                           // 设置字号大小
            paint.setColor(Color.GRAY);                      // 设置画笔颜色为灰色
            canvas.drawText("ANDROID",45,260,paint);         // 绘制文字
            super.onDraw(canvas);
        }
    }
```

然后，在布局管理器中使用定义的 MyView 控件：

```
<?xml version="1.0" encoding="utf-8"?>
<RelativeLayout xmlns:android="http://schemas.android.com/
                              apk/res/android"
    xmlns:tools="http://schemas.android.com/tools"
    xmlns:app=http://schemas.android.com/apk/res-auto
    android:layout_width="match_parent"
    android:layout_height="match_parent"
    android:paddingLeft="@dimen/activity_horizontal_margin"
    android:paddingRight="@dimen/activity_horizontal_margin"
    android:paddingTop="@dimen/activity_vertical_margin"
    android:paddingBottom="@dimen/activity_vertical_margin"
    app:layout_behavior="@string/appbar_scrolling_view_behavior"
    tools:showIn="@layout/activity_main"
    tools:context=".MainActivity"><com.example.canvaspaint
        android:layout_width="fill_parent"
        android:layout_height="fill_parent" />
</RelativeLayout>
```

程序在 Genymotion 中的运行效果如图 9.1 所示。

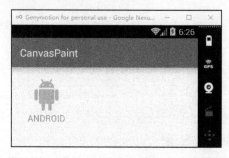

图 9.1　Canvas 类与 Paint 类实例运行效果

9.2 位图操作

在学会如何绘制图形后，懂得图片处理也是至关重要的。Bitmap 类，即位图，是 Android 中进行图片处理的基础且重要的类之一。使用该类可以实现图片文件信息的读取和写入，进行图片剪切、旋转、缩放等操作，并按指定格式保存图片文件。Bitmap 类的常用方法如表 9.3 所示。

表 9.3 Bitmap 类的常用方法

返回类型	方法	简述
static Bitmap	creatBitmap(Bitmap source, int x, int y, int width, int height)	从源位图中创建新的 Bitmap 对象
	creatScaledBitmap(Bitmap src, int dstWidth, int dstheight, boolean filter)	对源位图进行缩放，构成新位图
	creatBitmap(int width, int height, Bitmap.Config config)	根据指定的数据、配置创建新位图
	creatBitmap(Bitmap source, int x, int y, int width, int height, Matrix m, boolean filter)	从源位图中创建新的 Bitmap 对象，并进行 Matrix 变换
final int	getHeight()	获取位图的高度
	getWidth()	获取位图的宽度

在 Android 中，Bitmap 类比较特别，因为 Bitmap 类的构造函数是私有的，并不能实例化，只能借助 BitmapFactory 类获取文件资源，从而得到一个实例。BitmapFactory 类的常用绘图方法如表 9.4 所示。

表 9.4 BitmapFactory 类的常用绘图方法

返回类型	方法	简述
static Bitmap	decodeByteArray(byte[] data, int offset, int length)	根据指定的数据文件解码 Bitmap
	decodeFile(String pathName)	根据指定的路径创建 Bitmap
	decodeFileDescriptor(FileDescriptor fd)	根据指定的文件说明符解码 Bitmap
	decodeResource(Resources res, int id)	根据指定的资源创建 Bitmap
	decodeStream(InputStream is)	根据指定的输入流解码 Bitmap

从以上的方法中不难看出，BitmapFactory 类相当于从一个指定的文件中获取并解析 Bitmap 类。一般有两种读取 Bitmap 类的方法。

方法一：以 R 文件的方式。我们一般把图片放在/res/drawable 目录下，所以在程序中直接通过该图片对应的资源号就可以获取图片对象，例如，res/drawable 目录下有图片文件 test.png，代码如下：

```
Bitmap bitmap = BitmapFactory.decodeResource(this.getContext().getResources,
R.drawable.test)
```

方法二：以文件流的方式。假设在 sdcard 目录下有图片文件 test.png，代码如下：

```
FileInputStream fis = new FileInputStream("/sdcard/test.png");
Bitmap bitmap = BitmapFactory.decodeStream(fis);
```

【例 9.2】 Bitmap 类获取图片实例

下面的例子展示了如何运用 Bitmap 类获取图片并用按钮切换显示，具体实现方法如下：

```java
package com.example.bitmap;
import android.graphics.Bitmap;
import android.graphics.BitmapFactory;
import android.support.v7.app.AppCompatActivity;
import android.os.Bundle;
import android.view.View;
import android.widget.Button;
import android.widget.ImageView;
import import MainActivity import AppCompatActivity {
    ImageView img = null;
    Button but = null;
    int count = 0;
    int[] imgRes = {R.drawable.leaf1,R.drawable.leaf2,R.drawable.leaf3};
    @Override
    protected void onCreate(Bundle savedInstanceState) {
        super.onCreate(savedInstanceState);
        setContentView(R.layout.activity_main);
        Bitmap bm = BitmapFactory.decodeResource(this.getResources(),
                imgRes[count]);
        img = (ImageView)findViewById(R.id.imageView);
        but = (Button)findViewById(R.id.but_next);
        img.setImageBitmap(bm);
        but.setOnClickListener(new View.OnClickListener() {
            @Override
            public void onClick(View v) {
                count++;
                if(count==3)
                {count=0;}
                Bitmap bm = BitmapFactory.decodeResource(MainActivity.
                        this.getResources(),imgRes[count]);
                img.setImageBitmap(bm);
            }
        });
    }
}
```

本程序将保存在 res/drawable 目录中的图片资源提取出来，保存在 Bitmap 类中，并使用按钮切换资源图片。

在布局文件 activity_main.xml 中定义相关的控件，代码如下：

```xml
<?xml version="1.0" encoding="utf-8"?>
<LinearLayout xmlns:android="http://schemas.android.com/
                    apk/res/android"
    xmlns:tools="http://schemas.android.com/tools"
    tools:context=".MainActivity"
    android:layout_width="match_parent"
    android:layout_height="match_parent"
    android:orientation="vertical">
    <ImageView
```

```xml
        android:id="@+id/imageView"
        android:layout_width="fill_parent"
        android:layout_height="200dp" />
    <Button
        android:id="@+id/but_next"
        android:layout_width="wrap_content"
        android:layout_height="wrap_content"
        android:textSize="10dp"
        android:text="@string/next_pic"
        android:layout_gravity="center_horizontal"
        />
</LinearLayout>
```

在布局文件中定义了 ImageView，用于显示图片，以及一个按钮，用于切换图片。程序在 Genymotion 中的运行效果如图 9.2 所示。

图 9.2　Bitmap 类获取图片实例运行效果

但是在运行上述程序时，会出现一些卡顿的现象，甚至会出现 OOM（Out of Memory，内存溢出）问题，这是因为图片较大。所以在实际运用时，一般会通过设置 BitmapFactory.Options 的一些参数来对图片加载进行一些优化，代码如下：

```
BitmapFactory.Options opt = new BitmapFactory.Options();
opt.inSampleSize = 2;                /*2 倍缩放图片*/
opt.inDither=false;                  /*不进行图片抖动处理*/
opt.inPreferredConfig=null;          /*设置让解码器以最佳方式解码*/
```

在设置了上述几个参数之后，图片切换就会十分流畅。

9.3　Android 中的动画

在 Android 3.0 之前，Android 支持以下两种动画模式。

（1）Frame Animation（逐帧动画）：可以将一系列图片依次显示，以模拟动画效果，就像 GIF 图片一样。

（2）Tween Animation（补间动画）：对图片进行基本动画处理，如渐变、旋转、伸缩及移动的操作。

之后,又添加了一种新的动画方式。

Property Animation(属性动画):利用对象的属性变化实现动画效果,如按钮的位置和大小发生改变。

这3种动画在SDK中分别被称为Drawable Animation、View Animation、Property Animation。本节将详细介绍这3种动画的运用方式。

9.3.1 Frame Animation

Frame Animation,即逐帧动画,通过一系列Drawable控制动画过程中的每张静态图片依次显示,利用人眼"视觉暂留"的原理来实现动画效果。在XML文件中可以对这种动画进行配置,并将文件保存在res/drawable目录中。配置逐帧动画非常简单,在<animation-list.../>元素中逐个配置<item>元素即可定义动画的各个帧,并指定各个帧的持续时间,相关的配置属性及描述如表9.5所示。

表9.5 逐帧动画的配置属性及描述

属性	说明	参数说明
oneshot	循环属性	当被设置为true时,动画只循环一次,否则一直循环
drawable	动画资源	工程目录下的动画资源文件
visible	动画是否初始可见	当被设置为true时,动画初始可见,否则动画初始不可见
duration	动画持续时间	时间以毫秒为单位

【例9.3】 利用逐帧动画实现缓冲圈效果的实例

首先,在res/drawable目录中定义my_frameanimation.xml文件,代码如下:

```xml
<?xml version="1.0" encoding="utf-8"?>
<animation-list
    xmlns:android="http://schemas.android.com/apk/res/android"
    android:oneshot="false"
    >
    <item
        android:drawable="@drawable/pic1"
        android:duration="200"
        />
    <item
        android:drawable="@drawable/pic2"
        android:duration="200"
        />
    <item
        android:drawable="@drawable/pic3"
        android:duration="200"
        />
    <item
        android:drawable="@drawable/pic4"
        android:duration="200"
        />
    <item
        android:drawable="@drawable/pic5"
        android:duration="200"
        />
```

```xml
    <item
        android:drawable="@drawable/pic6"
        android:duration="200"
        />
    <item
        android:drawable="@drawable/pic7"
        android:duration="200"
        />
    <item
        android:drawable="@drawable/pic8"
        android:duration="200"
        />
</animation-list>
```

然后，在布局文件中添加 ImageView，代码如下：

```xml
<LinearLayout xmlns:android="http://schemas.android.com/apk/res/android"
    xmlns:tools="http://schemas.android.com/tools"
    android:id="@+id/linear"
    android:layout_width="fill_parent"
    android:layout_height="fill_parent"
    android:orientation="vertical" >
    <ImageView
        android:id="@+id/imageView1"
        android:layout_width="wrap_content"
        android:layout_height="wrap_content"
        android:layout_gravity="center" />
</LinearLayout>
```

最后，在程序中调用该逐帧动画，代码如下：

```java
package com.example.frameanimation;
import com.example.frameanimation.R;
import android.app.Activity;
import android.graphics.drawable.AnimationDrawable;
import android.os.Bundle;
import android.widget.ImageView;
public class MainActivity extends Activity {
    private ImageView mImg;
    private AnimationDrawable mAd;
    @Override
    protected void onCreate(Bundle savedInstanceState) {
        super.onCreate(savedInstanceState);
        setContentView(R.layout.activity_main);
        mImg = (ImageView) findViewById(R.id.imageView1);
        mImg.setBackgroundResource(R.drawable.my_frameanimation);
        mAd = (AnimationDrawable) MainActivity.this.mImg.getBackground();
        mAd.start();
    }
}
```

逐帧动画效果如图 9.3 所示。

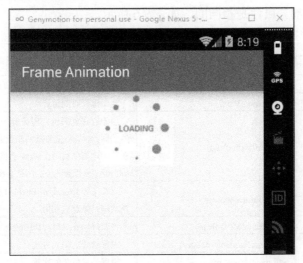

图 9.3　逐帧动画效果

9.3.2　Tween Animation

Tween Animation，即补间动画，可以让某个控件出现旋转、渐变、移动、缩放效果。补间动画的实现原理比较简单，开发者不需要定义每一帧的图像，只需要定义动画开始与结束的关键帧（首帧和尾帧）的图像，同时指定动画执行的时间，如图 9.4 所示。

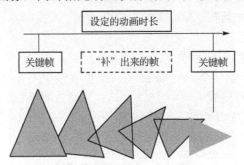

图 9.4　补间动画的实现原理

补间动画只能应用于 View 对象，而且只支持一部分属性，如支持缩放、旋转，而不支持背景、颜色的改变等。

补间动画主要由以下 4 种类型组成。

（1）alpha(android.view.animation.Alpha.Animation)：渐变动画，可实现控件的渐显和渐隐效果。

（2）rotate(android.view.animation.Rotate.Animation)：旋转动画，可实现控件的旋转效果。

（3）scale(android.view.animation.Scale.Animation)：缩放动画，可实现动态调整控件尺寸的效果。

（4）translate（android.view.animation.Translate.Animation)：位移动画，可实现对控件进行水平、垂直位移的效果。

Android 中的 4 种补间动画类型在 Java 中对应的类为 AlphaAnimation、ScaleAnimation、

TranslateAnimation 及 RotateAnimation。我们可以调用这些类的相关方法来获取和操作相应的属性。这 4 种类型的部分属性是相同的，如表 9.6 所示。

表 9.6　4 种补间动画类型的共同属性及描述

属　性	方　法	简　述
android:detachWallpaper	setDetachWallpaper(boolean)	当窗口在墙纸顶层时，窗口动，墙纸不动
android:duration	setDuration(long)	动画运行时间总计
android:fillAfter	setFillAfter(boolean)	由于存在动画链，假设有一个移动的动画紧跟一个淡出的动画，如果不把移动的动画的 setFillAfter 设置为 true，则在移动动画结束后，View 会回到原来的位置淡出；如果把 setFillAfter 设置为 true，则会在移动动画结束的位置淡出
android:fillBefore	setFillBefore(boolean)	当其值为 true 或 fillEnabled 的值为 false 时，该动画转化在动画开始前被应用
android:fillEnabled	setFillEnabled(boolean)	当其值为 true 时，fillBefore 的值被考虑
android:interpolator	setInterpolator(Interpolator)	设置动画的切入方式
android:repeatCount	setRepeatCount(int)	设置动画重复的次数
android:repeatMode	setRepeatMode(int)	设置动画重复的模式
android:startOffset	setStartOffset(long)	设置在动画运行前延迟数毫秒
android:zAdjustment	setZAdjustment(int)	设置动画的 ZOrder 的改变，0 表示不变；1 表示在最上层；-1 表示在最下层

可以通过编码实现动画，也可以使用 XML 文件形式定义动画。在通常情况下，使用 XML 文件形式定义动画，并将该文件放在/res/anim 目录下，这符合 MVC 开发规范，可以在需要时直接调用资源。下面介绍如何在 XML 文件中定义动画。

（1）新建 Android 项目。
（2）在 res 目录中新建 anim 目录。
（3）在 anim 目录中新建一个 new_anim.xml 文件（注意：在 Android 项目中，自建 XML 类型文件名必须都采用小写格式）。
（4）在 new_anim.xml 文件中加入动画代码。
① alpha（渐变透明度动画效果）属性及描述如表 9.7 所示。

表 9.7　alpha 属性及描述

属　性	说　明	参数说明
fromAlpha	动画起始时的透明度	取值为 0.0~1.0 的 float 数据类型的数字，0.0 表示完全透明，1.0 表示完全不透明
toAlpha	动画结束时的透明度	
duration	动画持续时间	时间以毫秒为单位

例如，实现一个持续时间为 1s、由 50%透明度到 100%透明度效果的动画，代码如下：

```xml
<?xml version="1.0" encoding="utf-8"?>
<set xmlns:android="http://schemas.android.com/apk/res/android" >
    <alpha
        android:duration="1000"
        android:fromAlpha="0.5"
        android:toAlpha="1.0" />
</set>
```

② rotate（画面移动旋转动画效果）属性及描述如表 9.8 所示。

表 9.8 rotate 属性及描述

属　　性	说　　明	参　数　说　明
fromDegrees	动画起始时控件的角度	当起始角度小于结束角度时，控件顺时针旋转
toDegrees	动画结束时控件旋转的角度	
pivotX	动画相对于控件的 X 坐标的开始位置	从 0%～100%中取值
PivotY	动画相对于控件的 Y 坐标的开始位置	50%表示控件的 X 或 Y 方向坐标的中点位置
duration	动画持续时间	时间以毫秒为单位

例如，实现一个持续时间为 1s、以控件中点为中心顺时针旋转 180°效果的动画，代码如下：

```xml
<?xml version="1.0" encoding="utf-8"?>
<set xmlns:android="http://schemas.android.com/apk/res/android" >
    <rotate
        android:duration="1000"
        android:fromDegrees="0"
        android:interpolator="@android:anim/accelerate_decelerate_
                    interpolator"
        android:pivotX="50%"
        android:pivotY="50%"
        android:toDegrees="+180" />
</set>
```

③ scale（渐变尺寸伸缩动画效果）属性及描述如表 9.9 所示。

表 9.9 scale 属性及描述

属　　性	说　　明	参　数　说　明
fromXScale	动画起始时 X 坐标的伸缩尺寸	0.0 表示收缩到没有
fromYScale	动画起始时 Y 坐标的伸缩尺寸	1.0 表示正常无伸缩
toXScale	动画结束时 X 坐标的伸缩尺寸	值小于 1.0 表示收缩
toYScale	动画结束时 Y 坐标的伸缩尺寸	值大于 1.0 表示放大
pivotX	动画相对于控件的 X 坐标的开始位置	从 0%～100%中取值
pivotY	动画相对于控件的 Y 坐标的开始位置	50%表示控件的 X 或 Y 方向坐标的中点位置
duration	动画持续时间	时间以毫秒为单位
fillAfter	当其值为 true 时，该动画转化在动画结束后被应用	

例如，实现一个持续时间为 1s、由正常无伸缩到 50%大小效果的动画，代码如下：

```xml
<?xml version="1.0" encoding="utf-8"?>
<set xmlns:android="http://schemas.android.com/apk/res/android" >
    <scale
        android:duration="1000"
        android:fillAfter="false"
        android:fromXScale="1.0"
        android:fromYScale="1.0"
        android:pivotX="50%"
        android:pivotY="50%"
        android:toXScale="0.5"
```

```
            android:toYScale="0.5" />
    </set>
```

④ translate（画面转换位置移动动画效果）属性及描述如表 9.10 所示。

表 9.10　translate 属性及描述

属　　性	说　　明	参 数 说 明
fromXDelta	动画起始时 X 坐标的位置	在没有指定的时候，默认以自己为相对参照物
fromYDelta	动画起始时 Y 坐标的位置	
toXDelta	动画结束时 X 坐标的位置	
toYDelta	动画结束时 Y 坐标的位置	
duration	动画持续时间	时间以毫秒为单位

例如，实现一个持续时间为 2s、具有移动效果的动画，代码如下：

```
<?xml version="1.0" encoding="utf-8"?>
<set xmlns:android="http://schemas.android.com/apk/res/android" >
    <translate
        android:duration="2000"
        android:fromXDelta="0"
        android:fromYDelta="0"
        android:toXDelta="180"
        android:toYDelta="180" />
</set>
```

XML 文件中必须至少有一个根元素，可以是<alpha><rotate><scale><translate>中的任意一个，也可以是<set>（一个由 4 个根元素中任意几个根元素组成的动画集合）。<set>是可以嵌套的。

有人曾形象地把<set>比喻为一个动画容器，当存在多个动画元素时，默认这些动画会同时发生，但也可以通过 startOffset 属性设置各个动画的开始时间，实现多个动画按顺序播放的效果。

使用上述方法可以决定动画的开始状态、结束状态及变化效果，通过设置动画持续时间，就可以计算出在中间插入多少帧的动画来实现动画效果。但是，在动画运行时，动画效果的变化速度还需要借助 Interpolator 来控制。Interpolator 可译为"插值器"，用于控制插入帧的属性，它可以依据特定的算法计算出整个动画所需要插入帧的密度与位置，从而使基本的动画实现匀速、加速、减速等各种变化速度的效果。Interpolator 接口的常用实现类如表 9.11 所示。

表 9.11　Interpolator 接口的常用实现类

方　　法	简　　述
AccelerateDecelerateInterpolator	动画开始、结束时减速，中间加速
AccelerateInterpolator	动画开始时速度较慢，然后开始加速
CycleInterpolator	动画循环播放指定次数，速度按正弦曲线改变
DecelerateInterpolator	动画开始时速度较快，然后开始减速
LinearInterpolator	动画以匀速变化

为了在动画资源中指定补间动画所使用的插值器，可以定义补间动画的<set>元素支持 android:interpolator 属性，并设置相应的参数，就可以控制动画效果的变化速度。Android 提供的 3 种动画插值器如表 9.12 所示。

表 9.12　Android 提供的 3 种动画插值器

属　性	简　述
accelerate_decelerate_interpolator	加速-减速动画插值器
accelerate_interpolator	加速动画插值器
decelerate_interpolator	减速动画插值器

【例 9.4】 补间动画实例

下面运用以上提供的各种动画属性实现一个动画切换的程序。

定义 Activity 程序：

```java
package com.example.tweenanimation;
import android.support.v7.app.AppCompatActivity;
import android.os.Bundle;
import android.view.View;
import android.view.animation.Animation;
import android.view.animation.AnimationUtils;
import android.widget.Button;
import android.widget.ImageView;
public class MainActivity extends AppCompatActivity {
    private Button btnAlpha,btnScale,btnTrans,btnRotate,btnAll;
    private ImageView mImg;
    private Animation mAnimation;
    @Override
    protected void onCreate(Bundle savedInstanceState) {
        super.onCreate(savedInstanceState);
        setContentView(R.layout.activity_main);
        btnAll = (Button) findViewById(R.id.btnAll);
        btnAlpha = (Button) findViewById(R.id.btnAlpha);
        btnScale = (Button) findViewById(R.id.btnScale);
        btnRotate = (Button) findViewById(R.id.btnRotate);
        btnTrans = (Button) findViewById(R.id.btnTrans);
        mImg = (ImageView) findViewById(R.id.imageView1);
        btnAlpha.setOnClickListener(new View.OnClickListener() {
            @Override
            public void onClick(View v) {
                mAnimation=AnimationUtils.loadAnimation(MainActivity.this,
                    R.anim.my_alpha);
                mImg.startAnimation(mAnimation);
            }
        });
        btnScale.setOnClickListener(new View.OnClickListener() {
            @Override
            public void onClick(View v) {
                mAnimation=AnimationUtils.loadAnimation(MainActivity.this,
                    R.anim.my_scale);
                mImg.startAnimation(mAnimation);
            }
        });
```

```java
        btnRotate.setOnClickListener(new View.OnClickListener() {
            @Override
            public void onClick(View v) {
                mAnimation = AnimationUtils.loadAnimation(MainActivity.this,
                        R.anim.my_rotate);
                mImg.startAnimation(mAnimation);
            }
        });
        btnTrans.setOnClickListener(new View.OnClickListener() {
            @Override
            public void onClick(View v) {
                mAnimation = AnimationUtils.loadAnimation(MainActivity.this,
                        R.anim.my_translate);
                mImg.startAnimation(mAnimation);
            }
        });
        btnAll.setOnClickListener(new View.OnClickListener() {
            @Override
            public void onClick(View v) {
                mAnimation = AnimationUtils.loadAnimation(MainActivity.this,
                        R.anim.my_all);
                mImg.startAnimation(mAnimation);
            }
        });
    }
}
```

其中的<alpha><rotate><scale><translate>在之前有所展示，这里不再重复。<set>同时包含多个动画元素的方法如下：

```xml
<?xml version="1.0" encoding="utf-8"?>
<set
    xmlns:android="http://schemas.android.com/apk/res/android" >
    <alpha
        android:duration="1000"
        android:fromAlpha="0.5"
        android:toAlpha="1.0" />
    <rotate
        android:duration="1000"
        android:fromDegrees="0"
        android:pivotX="50%"
        android:pivotY="50%"
        android:toDegrees="+180" />
    <scale
        android:duration="1000"
        android:fillAfter="false"
        android:fromXScale="1.0"
        android:fromYScale="1.0"
        android:pivotX="50%"
        android:pivotY="50%"
```

```
            android:toXScale="0.5"
            android:toYScale="0.5" />
    <translate
        android:duration="1000"
        android:fromXDelta="0"
        android:fromYDelta="0"
        android:toXDelta="180"
        android:toYDelta="180"/>
</set>
```

定义布局文件，代码如下：

```
<LinearLayout xmlns:android="http://schemas.android.com/apk/res/android"
    xmlns:tools="http://schemas.android.com/tools"
    android:layout_width="fill_parent"
    android:layout_height="fill_parent"
    android:orientation="vertical" >
    <ImageView
        android:id="@+id/imageView1"
        android:layout_width="200dp"
        android:layout_height="200dp"
        android:layout_gravity="center_horizontal"
        android:src="@drawable/pic" />
    <LinearLayout
        android:layout_width="fill_parent"
        android:layout_height="wrap_content"
        >
        <Button
            android:id="@+id/btnAlpha"
            android:layout_width="wrap_content"
            android:layout_height="wrap_content"
            android:text="ALPHA" />
        <Button
            android:id="@+id/btnRotate"
            android:layout_width="wrap_content"
            android:layout_height="wrap_content"
            android:text="ROTATE" />
        <Button
            android:id="@+id/btnScale"
            android:layout_width="wrap_content"
            android:layout_height="wrap_content"
            android:text="SCALE" />
        <Button
            android:id="@+id/btnTrans"
            android:layout_width="wrap_content"
            android:layout_height="wrap_content"
            android:text="TRANSLATE" />
    </LinearLayout>
    <Button
        android:id="@+id/btnAll"
```

```
            android:layout_width="wrap_content"
            android:layout_height="wrap_content"
            android:text="ALL" />
</LinearLayout>
```

补间动画效果如图 9.5 所示。

图 9.5 补间动画效果

9.3.3 Property Animation

Property Animation，即属性动画。与补间动画不同的是，属性动画会更改对象的实际属性，补间动画会改变 View 的绘制效果，而不改变 View 的属性。

属性动画的动画操作属性如表 9.13 所示。

表 9.13 属性动画的动画操作属性

属　　性	简述
Duration	动画持续时间，默认为 300ms
TimeInterpolation	定义动画的变化速度
Repeat count and behavior	定义重复次数、重复模式
Animator sets	动画集合
Frame refresh delay	帧刷新延迟

9.3.4 AnimationListener（动画监听器）

AnimationListener，即动画监听器，用于接收动画发出的表明其相关事件的通知，如动画已结束或动画被重复播放等。在 Android 系统中，有一个特定的接口用来监听动画的操作状态，即 android.view.animation.Animation.AnimationListener，其定义方法如表 9.14 所示。

表 9.14　AnimationListener 接口的定义方法

返回类型	方　　法	简　　述
abstract void	onAnimationCancel(Animator animation)	当动画被取消播放时被调用
	onAnimationEnd(Animator animation)	当动画结束时被调用
	onAnimationRepeat(Animator animation)	当动画被重复播放时被调用
	onAnimationStart(Animator animation)	当动画开始时被调用

9.4　在 Android 中播放音频与视频

智能手机能够播放音乐、视频文件已经成为不可或缺的多媒体功能。Android 在其多媒体框架中支持多种常见的媒体类型，所以开发者能够轻松地将音频、视频及图像集成到应用程序中；用户能够通过多种形式播放音/视频文件，可以利用 MediaPlayer 类提供的 API 直接读取应用中的资源文件（raw 文件）、文件系统中的单独文件，也可以从互联网上获取数据流。

9.4.1　MediaPlayer 介绍

在介绍 MediaPlayer 类之前，先介绍 MediaPlayer 的生命周期，如图 9.6 所示。该图清晰地描述了 MediaPlayer 的各个状态，也列举了主要方法的调用时序，每种方法只能在一些特定的状态下使用。

（1）Idle 状态：当使用 new()方法创建了一个 MediaPlayer 对象或者调用了其 reset()方法时，该 MediaPlayer 对象处于 Idle 状态。

（2）End 状态：通过 release()方法可以进入 End 状态。只要 MediaPlayer 对象不再被使用，就应当尽快将其通过 release()方法释放，以释放相关的软硬件资源，并且在进入 End 状态后，将不会再进入任何其他的状态了。

（3）Initialized 状态：当 MediaPlayer 调用 setDataSource()方法进入 Initialized 状态时，表示此时要播放的文件已经设置好了。

（4）Prepared 状态：在初始化之后，需要调用 prepare()方法（同步）或 prepareAsync()方法（异步）进入此状态，即预播放状态，表示目前的媒体文件没有任何问题，可以进行文件播放。

（5）Started 状态：一旦准备好 MediaPlayer，就可以调用 start()方法，使文件进入 Started 状态。可以使用 isPlaying()方法检测 MediaPlayer 是否处于 Started 状态。在该状态下，可以使用 seekTo()方法指定媒体播放位置。

（6）Paused 状态：在 Started 状态下调用 pause()方法可以暂停媒体播放，从而进入 Paused 状态，再使用 start()方法可以继续媒体播放，在 Paused 状态下仍然可以使用 seekTo()方法。

（7）Stopped 状态：在 Started 状态或 Paused 状态下可调用 stop()方法停止媒体播放，而想要重新播放处于 Stopped 状态的媒体文件，需要通过 prepareAsync()和 prepare()方法进入 Prepared 状态。

（8）PlaybackCompleted 状态：当媒体文件正常播放完毕，并且没有设置循环播放时，就会进入此状态，并触发 OnCompletionListener 的 onCompletion()方法。可以通过 stop()方法停止播放，也可以通过 start()方法重新播放，或者使用 seekTo()方法来重新定位播放位置。

（9）Error 状态：当用户在播放操作中出现错误时，会触发 OnErrorListener.onError()事件。此时媒体播放进入 Error 状态，可以通过 reset()方法重新返回 Idle 状态。

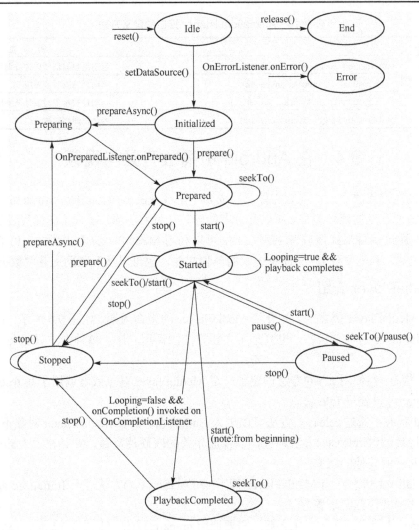

图 9.6 MediaPlayer 的生命周期

MediaPlayer 类的常用方法如表 9.15 所示。

表 9.15 MediaPlayer 类的常用方法

返回类型	方法	简述
boolean	isLooping()	检查 MediaPlayer 是否循环播放
	isPlaying()	检查 MediaPlayer 是否正在播放
int	getAudioSessionId()	返回音频的会话 ID
	getCurrentPosition()	获取当前播放的位置点
	getDuration()	获取文件的长度
	getVideoHeight()	返回视频的高度
	getVideoWidth()	返回视频的宽度
static	create(Context context, int resid)	从指定的源 ID 中创建一个 MediaPlayer 对象
	create(Context context, Uri uri)	从指定的 URI 中创建一个 MediaPlayer 对象
	create(Context context, Uri uri, SurfaceHolder holder)	从指定的 URI 中创建一个 MediaPlayer 对象，并在 SurfaceView 中显示

续表

返回类型	方法	简述
void	pause()	暂停播放
	prepare()	播放前同步准备播放器
	prepareAsync()	播放前异步准备播放器
	release()	释放此 MediaPlayer 对象占用的资源
	reset()	重置 MediaPlayer 到未初始化状态
	seekTo(int msec)	寻找特定的播放点
	selectTrack(int index)	选择路径
	setAudioAttributes(AudioAttributes attributes)	设置此 MediaPlayer 的音频属性
	setAudioSessionId(int sessionId)	设置音频的会话 ID
	setAudioStreamType(int streamtype)	设置音频类型
	setDataSource(String path)	设置数据文件（file-path 或 http/rtsp URL）
	setDataSource(Context context, Uri uri)	指定数据源
	setDisplay(SurfaceHolder sh)	设置视频显示
	setLooping(boolean looping)	设置循环
	setOnCompletionListener(MediaPlayer.OnCompletionListener listener)	当媒体播放完成后被触发
	setOnErrorListener(MediaPlayer.OnErrorListener listener)	当出现错误时被触发
	setOnPreparedListener(MediaPlayer.OnPreparedListener listener)	当媒体准备完成后被触发
	setOnSeekCompleteListener(MediaPlayer.OnSeekCompleteListener listener)	当媒体播放点设置完成后被触发
	setOnVideoSizeChangedListener(MediaPlayer.OnVideoSizeChangedListener listener)	当视频文件大小改变后被触发
	setVideoScalingMode(int mode)	设置视频缩放比例
	setVolume(float leftVolume, float rightVolume)	设置播放音量
	start()	开始或恢复播放
	stop()	停止播放

9.4.2 运用 MediaPlayer 播放音频

在熟悉了 MediaPlayer 类的常用方法之后，下面展示如何运用 MediaPlayer 播放一个 MP3 格式的音频。本例中使用了"播放""暂停""停止""重播"4 个按钮来控制音频的播放，并且使用了一个 SeekBar 来表示播放的进度，用户可以通过拖动进度条设置播放位置。需要播放的音频文件为 my_music.mp3，该文件被保存在 res/raw/目录下。

【例 9.5】 运用 MediaPlayer 播放 MP3 格式的音频实例

首先定义布局文件，代码如下：

```
<?xml version="1.0" encoding="utf-8"?>
<LinearLayout xmlns:android="http://schemas.android.com/apk/res/android"
    xmlns:tools="http://schemas.android.com/tools"
    android:layout_width="match_parent"
    android:layout_height="match_parent"
    android:orientation="vertical"
```

```xml
<LinearLayout
    android:layout_width="wrap_content"
    android:layout_height="wrap_content"
    android:orientation="horizontal">
    <TextView
        android:layout_width="wrap_content"
        android:layout_height="wrap_content"
        android:text="@string/play_state"
        android:textSize="25sp"
        />
    <TextView
        android:id="@+id/tvState"
        android:layout_width="wrap_content"
        android:layout_height="wrap_content"
        android:text="@string/stete_prepare"
        android:textSize="25sp"
        />
</LinearLayout>

<LinearLayout
    android:layout_width="match_parent"
    android:layout_height="wrap_content"
    android:layout_marginTop="10dp"
    android:orientation="horizontal">
    <Button
        android:id="@+id/btnStart"
        android:layout_width="wrap_content"
        android:layout_height="wrap_content"
        android:layout_marginLeft="5dp"
        android:text="@string/start"/>
    <Button
        android:id="@+id/btnPause"
        android:layout_width="wrap_content"
        android:layout_height="wrap_content"
        android:layout_marginLeft="5dp"
        android:text="@string/puase"/>
    <Button
        android:id="@+id/btnEnd"
        android:layout_width="wrap_content"
        android:layout_height="wrap_content"
        android:layout_marginLeft="5dp"
        android:text="@string/end"/>
    <Button
        android:id="@+id/btnRestart"
        android:layout_width="wrap_content"
        android:layout_height="wrap_content"
        android:layout_marginLeft="5dp"
        android:text="@string/restart"/>
</LinearLayout>
```

```xml
<SeekBar
    android:layout_width="match_parent"
    android:layout_height="wrap_content"
    android:layout_marginTop="10dp"
    android:id="@+id/seekBar"
    android:layout_gravity="center_horizontal"
    android:indeterminate="false" />
</LinearLayout>
```

然后定义 Activity 程序。在 Activity 程序中，主要对其中的几个按钮及 SeekBar 定义监听；其次对 SeekBar 进行控制，以及运用一个异步线程刷新音频的播放情况。代码如下：

```java
package com.example.mediaplayer;
import android.app.Activity;
import android.media.MediaPlayer;
import android.os.AsyncTask;
import android.os.Bundle;
import android.view.View;
import android.view.View.OnClickListener;
import android.widget.Button;
import android.widget.SeekBar;
import android.widget.TextView;
public class MainActivity extends Activity implements OnClickListener{
    private MediaPlayer mediaPlayer;
    private TextView tvState;
    private Button btnPlay,btnPause,btnStop,btnRestart;
    private SeekBar seekBar;
    private boolean playFlag = false,pauseFlag = false;
    @Override
    protected void onCreate(Bundle savedInstanceState) {
        super.onCreate(savedInstanceState);
        setContentView(R.layout.activity_main);
        btnPlay = (Button)this.findViewById(R.id.btnStart);
        btnPause = (Button) this.findViewById(R.id.btnPause);
        btnStop = (Button) this.findViewById(R.id.btnEnd);
        btnRestart= (Button) this.findViewById(R.id.btnRestart);
        tvState= (TextView)this.findViewById(R.id.tvState);
        seekBar = (SeekBar)this.findViewById(R.id.seekBar);
        btnPlay.setOnClickListener(this);
        btnPause.setOnClickListener(this);
        btnStop.setOnClickListener(this);
        btnRestart.setOnClickListener(this);
        mediaPlayer = MediaPlayer.create(MainActivity.this, R.raw.my_music);
    }

    /**
    *在按键监听函数中运用 start()、stop()、pause()、release()方法实现对播放的控制
    */
    @Override
    public void onClick(View v) {
    switch (v.getId())
```

```java
        {
            case R.id.btnStart:
                if(pauseFlag==false) {
                    mediaPlayer.setOnCompletionListener(new MediaPlayer
                      .OnCompletionListener() {
                        @Override
                        public void onCompletion(MediaPlayer mp) {
                            tvState.setText(getResources().getString(R.string.stete_end));
                            playFlag = false;
                            mp.release();
                        }
                    });
                    seekBar.setMax(mediaPlayer.getDuration());
                    playFlag = true;
                    UpdateSeekBar updateProgress = new UpdateSeekBar();
                    updateProgress.execute(1000);
                    seekBar.setOnSeekBarChangeListener(new
                        MySeekBarChangeListener());
                    try {
                        if (mediaPlayer != null) {
                            mediaPlayer.stop();
                        }
                            mediaPlayer.prepare();
                            mediaPlayer.start();
                            tvState.setText(getResources().getString(R.string.start));
                        } catch (Exception e) {
                     this.tvState.setText(getResources()
                      .getString(R.string.stete_error));
                        }
                }else {
                        pauseFlag = false;
                        playFlag = true;
                        mediaPlayer.start();
                        UpdateSeekBar updateProgress = new UpdateSeekBar();
                        updateProgress.execute(1000);
                        seekBar.setOnSeekBarChangeListener(new
                            MySeekBarChangeListener());
                        tvState.setText(getResources().getString(R.string.start));
                    }
                break;
            case R.id.btnEnd:
                if(mediaPlayer.isPlaying()){
                    mediaPlayer.stop();
                    playFlag = false;
                    tvState.setText(getResources().getString(R.string.end));
                }
                break;
            case R.id.btnPause:
                if(mediaPlayer.isPlaying())
                {
                    playFlag = false;
```

```java
                    pauseFlag=true;
                    mediaPlayer.pause();
                    tvState.setText(getResources().getString(R.string.puase));
                }
                break;
            case R.id.btnRestart:
                if(mediaPlayer.isPlaying()){
                    playFlag = true;
                    mediaPlayer.seekTo(0);
                }
                break;
        }
    }
    /**
    *在异步线程中实现对音频播放进度的显示
    */
    private class UpdateSeekBar extends AsyncTask<Integer,Integer,String> {
        @Override
        protected void onPostExecute(String s) {
        }
        @Override
        protected void onProgressUpdate(Integer... values) {
            seekBar.setProgress(values[0]);
        }
        @Override
        protected String doInBackground(Integer... params) {
            while(playFlag){
                try {
                    Thread.sleep(params[0]);
                } catch (InterruptedException e) {
                    e.printStackTrace();
                }
                this.publishProgress(mediaPlayer.getCurrentPosition());
            }
            return null;
        }
    }
    /**
    *在SeekBar监听中实现对音频播放的控制
    */
    private class MySeekBarChangeListener implements SeekBar.
            OnSeekBarChangeListener {
        @Override
        public void onProgressChanged(SeekBar seekBar, int progress,
                boolean fromUser) {
        }
        @Override
        public void onStartTrackingTouch(SeekBar seekBar) {
        }
        @Override
        public void onStopTrackingTouch(SeekBar seekBar) {
```

```
            mediaPlayer.seekTo(seekBar.getProgress());
        }
    }
}
```

程序在 Genymotion 中的运行效果如图 9.7 所示。

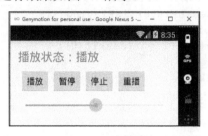

图 9.7 运行效果

9.4.3 播放视频

在 Android 中播放视频有 3 种方式。
- 使用 Android 自带播放器播放视频。
- 使用 VideoView 播放视频。
- 使用 MediaPlayer 和 SurfaceView 播放视频。

下面介绍使用以上 3 种方式播放一段视频的方法，并将文件放置在 SD 卡的/sdcard/Download 目录下。在本实例中，向模拟器中导入视频的方式有以下两种。

方式一：打开 Android Device Monitor，向模拟器的文件中导入视频文件，如图 9.8 所示。

图 9.8 打开 Android Device Monitor

选择"File Explorer"选项卡并单击右侧的按钮以选择相关的文件，就能将其导入虚拟机的/sdcard/Download 目录下，如图 9.9 所示。

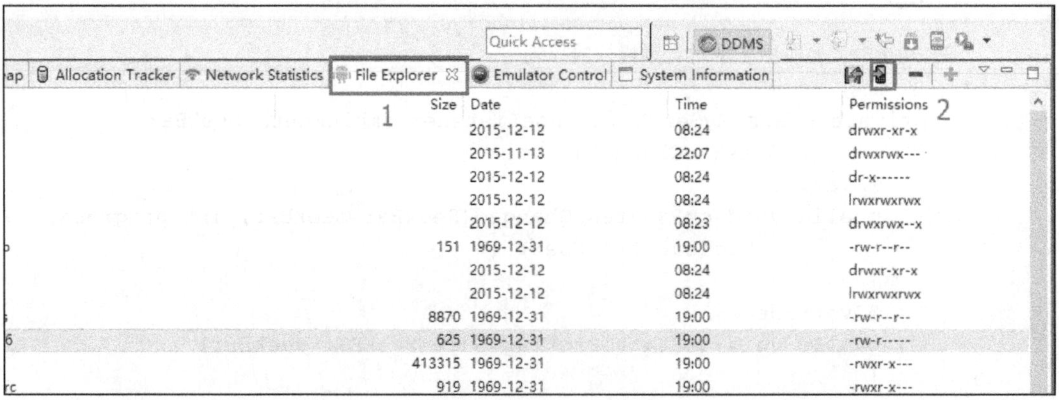

图 9.9 导入文件

第 9 章　图形、图片与多媒体

方式二：打开虚拟机后，可直接将文件拖动到虚拟机中的 /sdcard/Download 目录下，即与方式一中相同的目录下。导入后的文件如图 9.10 所示。

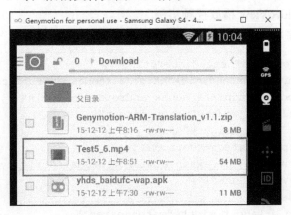

图 9.10　导入后的文件

注意：在调用 SD 卡中的文件时需要声明权限，不然会出现无法播放的情况。在工程项目的 AndroidManifest.xml 文件中加入以下代码，实现对 SD 卡的读/写操作：

```
<uses-permission android:name=
"android.permission.WRITE_EXTERNAL_STORAGE"/>
<uses-permission android:name=
"android.permission.MOUNT_UNMOUNT_FILESYSTEMS"/>
```

（1）使用 Android 自带播放器播放视频，指定 Action 为 ACTION_VIEW，Data 为 URI。

【例 9.6】　使用 Android 自带播放器播放视频实例

在 Activity 程序中定义如下代码：

```
package com.example.videoplayer;
import android.content.Intent;
import android.net.Uri;
import android.os.Environment;
import android.support.v7.app.AppCompatActivity;
import android.os.Bundle;
public class MainActivity extends AppCompatActivity {
    @Override
    protected void onCreate(Bundle savedInstanceState) {
        super.onCreate(savedInstanceState);
        setContentView(R.layout.activity_main);
        Uri uri = Uri.parse(Environment.getExternalStorageDirectory()
                .getPath()+"/Download/Test5_6.mp4");
        Intent it = new Intent(Intent.ACTION_VIEW);
        it.setDataAndType(uri,"video/mp4");
        startActivity(it);
    }
}
```

在执行该程序之后，会立即打开一个跳转页面，用户可在此选择播放器。

（2）使用 VideoView 播放视频。这个控件使用起来十分方便，首先在布局文件中创建该控件，然后在 Activity 中获取该控件，接下来通过 setVideoURI()方法或 setVideoPath()方法加载要播放的视频文件，最后使用 start()方法播放视频。

【例 9.7】 使用 VideoView 播放视频实例

首先定义布局文件，代码如下：

```xml
<?xml version="1.0" encoding="utf-8"?>
<LinearLayout xmlns:android="http://schemas.android.com/apk/res/android"
    xmlns:tools="http://schemas.android.com/tools"
    android:layout_width="match_parent"
    android:layout_height="match_parent"
    android:orientation="vertical"
    tools:context="com.example.videoview.MainActivity"
    android:weightSum="1">
    <VideoView
        android:layout_width="match_parent"
        android:layout_height="match_parent"
        android:id="@+id/videoView" />
</LinearLayout>
```

然后定义 Activity 程序，代码如下：

```java
package com.example.videoview;
import android.net.Uri;
import android.os.Environment;
import android.support.v7.app.AppCompatActivity;
import android.os.Bundle;
import android.widget.MediaController;
import android.widget.VideoView;
public class MainActivity extends AppCompatActivity {
    VideoView myVideo;
    @Override
    protected void onCreate(Bundle savedInstanceState) {
        super.onCreate(savedInstanceState);
        setContentView(R.layout.activity_main);
        MediaController myMc = new MediaController(MainActivity.this);
        Uri uri = Uri.parse(Environment.getExternalStorageDirectory()
                .getPath()+"/Download/Test5_6.mp4");
        myVideo = (VideoView)this.findViewById(R.id.videoView);
        myVideo.setVideoURI(uri);
        myVideo.setMediaController(myMc);
        myVideo.requestFocus();
        try{
            myVideo.start();
        }catch (Exception e){
            e.printStackTrace();
        }
    }
}
```

虽然 VideoView 在使用的过程中十分方便，但是由于 VideoView 在视频格式上有一定的条件限制，只有少数满足其格式要求的视频文件才能被完美地播放，因此在实际运用中不推荐使用这种方式。

（3）使用 MediaPlayer 和 SurfaceView 播放视频。

在上一节中展示了如何运用 MediaPlayer 播放音频，同样地，MediaPlayer 也被用于播放视频，但是仅使用 MediaPlayer 是不够的，因为它不能提供视频播放的界面。这时，就需要使用 SurfaceView 创建视频播放的界面。SurfaceView 能够提供快速的 GUI 更新，并且启动一个新的线程来处理视频播放中的图形加载，使得视频播放更为流畅。如果要设置播放界面的大小，则还需要 SurfaceHolder 类来提供某些方法。表 9.16 展示了 SurfaceHolder 类定义的相关方法。

表 9.16 SurfaceHolder 类定义的相关方法

返回类型	方法	简述
abstract Surface	getSurface()	界面对象的直接接口
abstract void	setFixedSize(int width, int height)	设置界面的固定尺寸
	setKeepScreenOn(boolean screenOn)	设置在界面显示时屏幕是打开的
	setSizeFromLayout()	允许界面基于其容器调整大小
	setType(int type)	设置界面的类型，一般会自动设置

【例 9.8】 使用 SurfaceView 与 MediaPlayer 播放视频实例

下面展示如何使用 SurfaceView 与 MediaPlayer 播放视频。

首先建立布局文件，代码如下：

```
<?xml version="1.0" encoding="utf-8"?>
<LinearLayout xmlns:android="http://schemas.android.com/apk/res/android"
    xmlns:tools="http://schemas.android.com/tools"
    android:layout_width="match_parent"
    android:layout_height="match_parent"
    android:paddingBottom="@dimen/activity_vertical_margin"
    android:paddingLeft="@dimen/activity_horizontal_margin"
    android:paddingRight="@dimen/activity_horizontal_margin"
    android:paddingTop="@dimen/activity_vertical_margin"
    android:orientation="vertical"
    tools:context="com.example.surfaceview.MainActivity" >
    <SurfaceView
        android:layout_width="match_parent"
        android:layout_height="200dp"
        android:id="@+id/mySurface"/>
    <Button
        android:layout_width="wrap_content"
        android:layout_height="wrap_content"
        android:text="@string/tv_play"
        android:id="@+id/btnStart"/>
    <Button
        android:layout_width="wrap_content"
        android:layout_height="wrap_content"
        android:text="@string/tv_pause"
        android:id="@+id/btnPause"/>
    <Button
```

```xml
        android:layout_width="wrap_content"
        android:layout_height="wrap_content"
        android:text="@string/tv_stop"
        android:id="@+id/btnStop"/>
</LinearLayout>
```

在 Activity 程序中，让 Activity 继承于 SurfaceHolder.Callback，使得在创建视频时复写其方法以实现自动加载。然后通过"播放""暂停""停止"3 个按钮控制视频的播放。具体实现代码如下：

```java
package com.example.surfaceview;
import android.app.Activity;
import android.media.AudioManager;
import android.media.MediaPlayer;
import android.net.Uri;
import android.os.Environment;
import android.os.Bundle;
import android.view.SurfaceHolder;
import android.view.SurfaceView;
import android.view.View;
import android.widget.Button;
public class MainActivity extends Activity implements SurfaceHolder.Callback{
    MediaPlayer mp;
    SurfaceView mySurface;
    SurfaceHolder surfaceHolder;
    Button btnPlay,btnPause,btnStop;
    @Override
    public void onCreate(Bundle savedInstanceState) {
        super.onCreate(savedInstanceState);
        setContentView(R.layout.activity_main);
        btnPlay=(Button)findViewById(R.id.btnStart);
        btnPause=(Button)findViewById(R.id.btnPause);
        btnStop=(Button)findViewById(R.id.btnStop);
        mySurface=(SurfaceView)findViewById(R.id.mySurface);
        mp = new MediaPlayer();
        // SurfaceHolder 是 SurfaceView 的控制接口
        surfaceHolder=mySurface.getHolder();
        surfaceHolder.setKeepScreenOn(true);
        surfaceHolder.setFixedSize(320,220);
        surfaceHolder.setType(SurfaceHolder
                    .SURFACE_TYPE_PUSH_BUFFERS);        // Surface 类型
        surfaceHolder.addCallback(this);
        btnPlay.setOnClickListener(new View.OnClickListener() {
            @Override
            public void onClick(View v) {
                play();
            }});
        btnPause.setOnClickListener(new View.OnClickListener(){
            @Override
```

```java
            public void onClick(View v) {
                mp.pause();
            }});
        btnStop.setOnClickListener(new View.OnClickListener(){
            @Override
            public void onClick(View v) {
                mp.stop();
            }});
    }
    @Override
    protected void onDestroy() {
        // TODO Auto-generated method stub
        super.onDestroy();
        if(mp.isPlaying()){
            mp.stop();
            // 在销毁Activity时停止播放，释放资源
            // 若不做这个操作，即使退出也能听到视频播放的声音
        }
        mp.release();
    }
    public void play(){
        try {
            mp.reset();
            mp.setAudioStreamType(AudioManager.STREAM_MUSIC);
            Uri uri = Uri.parse(Environment.getExternalStorageDirectory()
                    .getPath()+"/Test1.mp4");
            mp.setDataSource(MainActivity.this,uri);
            mp.setDisplay(mySurface.getHolder());
            mp.prepare();
            mp.start();
        }
        catch (Exception e){
            e.printStackTrace();
        }
    }
    @Override
    public void surfaceCreated(SurfaceHolder holder) {
        try {
            play();
        } catch (Exception e) {
            // TODO Auto-generated catch block
            e.printStackTrace();
        }
    }

    @Override
    public void surfaceChanged(SurfaceHolder holder, int format,
        int width, int height) {
    }
```

```
@Override
public void surfaceDestroyed(SurfaceHolder holder) {
}
}
```

视频播放效果如图 9.11 所示。

图 9.11　视频播放效果

9.5　控制摄像头拍照

Camera 类可用于设置图像捕捉的相关参数、启动/停止预览、拍照和检索视频的帧的编码，相当于一个 Camera 服务的客户端，可管理实际的摄像头硬件。

在访问设备的相机之前，必须在 Android Manifest 中声明 CAMERA 权限。当然，还要包括<uses-feature>manifest 元素所声明的摄像头功能由用户的应用程序使用。比如，使用摄像头和自动对焦功能，Android Manifest 应包括如下内容：

```
<uses-permission android:name="android.permission.CAMERA" />
<uses-feature android:name="android.hardware.camera" />
<uses-feature android:name="android.hardware.camera.autofocus" />
```

使用 Camera 类控制摄像头拍照的步骤如下。
（1）从 open(int)方法中获取一个摄像头的实例。
（2）在 getParameters()方法中得到默认设置参数。
（3）如果有必要，可修改返回的 Camera.Parameters 对象并调用，代码如下：

```
setParameters(Camera.Parameters)
```

（4）如果需要，可调用 setDisplayOrientation(int)方法。
（5）将一个完全初始化的 SurfaceHolder 传递给 setPreviewDisplay(SurfaceHolder)方法十分重要。否则，相机将无法启动预览。
（6）调用 startPreview()方法开始更新预览表。在预览开始之前，必须拍摄一张照片。
（7）可以调用 takePicture(Camera.ShutterCallback,Camera.PictureCallback,Camera.PictureCallback,

Camera.PictureCallback)方法捕捉照片，并等待回调函数提供实际的图像数据。

（8）在拍照后，将停止预览显示。如果需要拍摄更多照片，可再次调用 startPreview()方法。

（9）调用 stopPreview()方法停止更新预览表。

（10）调用 release()方法释放由其他应用程序使用的摄像头。应用程序应该在调用 onPause()方法后立即释放摄像头。

Camera 类的常用方法如表 9.17 所示。

表 9.17　Camera 类的常用方法

返回类型	方　　法	简　　述
Camera.Parameters	getParameters()	返回摄像头的参数
final void	autoFocus(Camera.AutoFocusCallback cb)	自动对焦，且当聚焦后记录一个回调函数
	cancelAutoFocus()	取消进程中的所有自动对焦功能
	lock()	锁定摄像头，阻止其他程序访问
	reconnect()	在其他程序调用摄像头后，重新连接摄像头服务
	release()	断开连接并释放摄像头资源
	startPreview()	开始捕捉、预览
	stopPreview()	停止捕捉、预览
	setDisplayOrientation(int degrees)	设置摄像头角度
	setZoomChangeListener(Camera.OnZoomChangeListener listener)	在显示区域发生改变时被触发
	takePicture(Camera.ShutterCallback shutter, Camera.PictureCallback raw, Camera.PictureCallback jpeg)	捕捉图像
	unlock()	解锁摄像头，允许其他程序访问
static int	getNumberOfCameras()	返回物理摄像头的个数
static void	getCameraInfo(int cameraId, Camera.CameraInfo cameraInfo)	返回指定摄像头的信息
static Camera	open(int cameraId)	创建新的 Camera 对象以打开指定的摄像头
	open()	创建新的 Camera 对象以打开默认的摄像头
void	setParameters(Camera.Parameters params)	设置摄像头的参数

SurfaceHolder.Callback 接口定义的方法如表 9.18 所示。

表 9.18　SurfaceHolder.Callback 接口定义的方法

返回类型	方　　法	简　　述
abstract void	surfaceChanged(SurfaceHolder holder, int format, int width, int height)	当界面的格式和大小发生改变时被调用
	surfaceCreated(SurfaceHolder holder)	当界面创建后被调用
	surfaceDestroyed(SurfaceHolder holder)	当界面关闭前被调用

Camera 类定义的接口如表 9.19 所示。

表 9.19　Camera 类定义的接口

方　　法	简　　述
Camera.AutoFocusCallback	自动对焦的回调接口
Camera.ErrorCallback	当出现错误时的回调接口

续表

方　　法	简　　述
Camera.FaceDetectionListener	当预览时检测到人脸的回调接口
Camera.OnZoomChangeListener	当显示区域改变时的回调接口
Camera.PictureCallback	当图片被捕捉后提供数据的回调接口
Camera.PreviewCallback	预览时的回调接口
Camera.ShutterCallback	标记实际捕捉图像操作的回调接口
SensorEventListener	当传感器参数改变时的回调接口

第 10 章 Android 网络编程基础

在 PC 时代,以太网为人们提供了极大的便利,各种信息服务使得计算机成为不可或缺的工具。随着移动终端的普及,移动互联网所带来的便捷性也不言而喻,网上购物、网络游戏、即时通信等都需要网络技术的支持。本章将由浅入深地讲解 Android 网络编程方法,从使用最基础的 TCP/IP 网络通信协议实现数据交流,到使用 WebService 技术访问网络数据,并开发一款手机归属地查询小应用。

10.1 基于 TCP 协议的网络通信

在网络环境中,由程序创建的"字节序列"被称为"分组报文"。一组报文不仅包括网络用来完成工作的控制信息,还包括数据内容。协议相当于互联网中相互通信的程序间的一种约定,规定了分组报文进行数据交换的方式与含义。

在网络层中,IP 协议用于完成将分组报文传输到指定目的地的任务,这时网络通信就完成了点对点(Port to Port)的传输,但并不能精确到程序。在传输层中,TCP 协议用于完成更细致的寻址操作,将报文发送到对应的程序中,这时网络通信就完成了端对端(End to End)的传输。

在网络连接中最基础的两个协议就是 TCP/IP 与 UDP,前者是面向连接的协议,能够为数据提供高效可靠的传输机制;而后者则是面向无连接的广播通信机制的协议,适用于数据量少、对通信质量要求不高的环境。在 Android 网络编程中,为了保证数据通信安全可靠,大多使用 TCP 协议进行数据交流。本节重点介绍基于 TCP 协议的网络通信方法。

10.1.1 TCP/IP 协议基础

TCP(Transmission Control Protocol,传输控制协议)是面向连接的通信协议。在使用互联网设备进行通信之前,数据交流的双方必须建立安全可靠的连接。这一过程类似于打电话,在拨通电话后,先说一声"喂",等对方确认了再开始正常通话。TCP 协议使用了重传机制,接收端在收到发送端发送的一个报文之后,会返回响应报文,如果发送端没有收到这个响应报文,则会继续重发,这样即使在网络拥塞时,也不会出现传输错误的情况。

在 Java 中,JDK 提供了基于 TCP、UDP 协议进行网络通信的 API,其中,Socket、ServerSocket 两个类用来建立基于 TCP/IP 协议的网络通信。Socket 本质上是对传输层中的 TCP 协议进行的封装,由于基于 TCP 协议建立的是端到端的通信,因此要实现 Socket 的传输,就需要构建客户端与服务端。基于 TCP 协议的网络通信原理如图 10.1 所示。

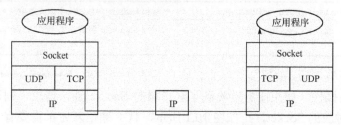

图 10.1 基于 TCP 协议的网络通信原理

10.1.2 使用 Socket 与 ServerSocket 建立通信

在介绍使用 Socket 与 ServerSocket 建立通信之前，需要了解数据通信中涉及的一些异常类型，这些异常类型在后文中将会用到。异常类型如表 10.1 所示。

表 10.1 异常类型

异常类型	描 述
UnkownHostException	主机名称或 IP 地址错误
ConnectException	服务器拒绝连接、服务器没有启动（超出队列数，拒绝连接）
SocketTimeoutException	连接超时
BindException	Socket 对象无法与指定的本地 IP 地址或端口绑定

在图 10.1 所示的基于 TCP 协议的网络通信原理中，由于两个通信实体已经建立连接，因此并没有区分客户端与服务端，但在两个通信实体通信之前，其中作为服务端的一方需要建立等待连接的机制，等待客户端的请求，这时就需要 ServerSocket 对象监听来自客户端的 Socket 连接，如果没有连接，则服务端将会一直处于等待状态。Socket 与 ServerSocket 的交互原理如图 10.2 所示。

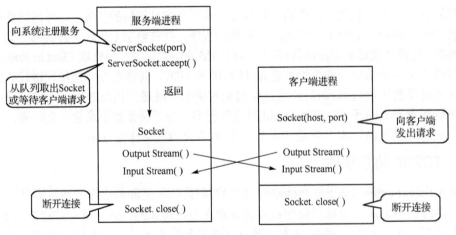

图 10.2 Socket 与 ServerSocket 的交互原理

由图 10.2 可知，要完成 Socket 通信，首先需要建立服务端，ServerSocket 类提供了如表 10.2 所示的构造方法，用于创建 ServerSocket 对象。

表 10.2 ServerSocket 类的构造方法

方 法	描 述
ServerSocket(int port) throws IOException	创建一个指定端口的 ServerSocket 对象，端口范围为 0~65 535
ServerSocket(int port, int backlog) throws IOException	增加 backlog 参数，用于改变连接队列长度
ServerSocket(int port, int backlog, InetAddress bindAddr) throws IOException	增加 bindAddr 参数，用于将 ServerSocket 对象绑定到指定的 IP 地址

在指定了服务端之后，就需要监听来自客户端的连接请求。ServerSocket 类提供了 accept() 方法用于执行上述操作。如果服务端收到了客户端的 Socket 连接请求，则该方法将返回一个与客户端 Socket 对应的 Socket，正如图 10.1 所示，TCP 通信双方都需要有一个 Socket。

在使用 ServerSocket 时需要注意以下几点。

第 10 章 Android 网络编程基础

- 如果端口被占用或者没有权限使用某些端口,则会抛出 BindException 错误。例如,在使用 1~1023 端口时,要求用户具有管理员权限。
- 如果将端口设置为 0,则系统会自动为其分配一个端口。
- ServerSocket 一旦绑定了监听端口,就无法更改。使用 ServerSocket()方法可以实现在绑定端口前设置其他的参数。

通常在使用完 ServerSocket 之后,可以调用 ServerSocket 类的 close()方法关闭该 ServerSocket。下面的代码展示了如何运用 ServerSocket 建立服务端:

```
ServerSocket serverSocket = new ServerSocket(8888);// 设置服务端的端口为 8888
while(true){
try{
    Socket socket=serverSocket.accept();
              // 从连接队列中取出一个连接,如果没有则等待
System.out.println("新增连接:"+socket.getInetAddress()+":"+socket.getPort());
…// 接收和发送数据
}catch(IOException e){e.printStackTrace();}
finally{
    try{
        if(socket!=null)
        socket.close();        // 在与一个客户端通信结束后,要关闭 Socket
    }catch(IOException e){e.printStackTrace();}
    }
}
```

注意:由于服务端通常运行在有固定 IP 地址的服务器上,因此上述代码可以直接运行在 PC 上,将 PC 当作服务端。

在建立好服务端之后,就可以开始建立客户端了,通常使用 Socket 类的构造方法直接指定需要连接的服务器 IP 地址。Socket 类提供了如表 10.3 所示的几个构造方法。

表 10.3 Socket 类的几个构造方法

方 法	描 述
Socket(InetAddress address/String host, int port)throws UnknownHostException, IOException	创建连接指定的远程 IP 地址及端口的 Socket,若未指定本地 IP 地址及端口,则采用本地主机的 IP 地址
Socket(InetAddress address/String host, int port, InetAddress localAddress/String host, int localPort)throws IOException	对比上面的方法,指定了本地主机的 IP 地址及端口

Socket 类还提供了一些常用于建立通信连接的方法,如表 10.4 所示。

表 10.4 Socket 类的常用方法

返回类型	方 法	描 述
InetAddress	getInetAddress()	获取远程服务端的 IP 地址
int	getPort()	获取远程服务端的端口
String	getLocalAddress()	获取本地客户端的 IP 地址
int	getLocalPort()	获取本地客户端的端口
InputString	getInputStream()	获取输入流
OutputString	getOutputStream()	获取输出流

续表

返回类型	方法	描述
void	connect(SocketAddress remoteAddr)	连接指定的 IP 地址
void	connect(SocketAddress remoteAddr, int timeout)	连接指定的 IP 地址，如果时间超过 timeout，则抛出连接超时异常

在上述方法中，getInputStream()方法与 getOutputStream()方法最为重要，用于在通信时获取输入流与输出流。下面的例子展示了如何运用 Socket 类建立通信连接，实现连接指定 IP 端口并获取服务端发送的数据的功能。该程序创建连接后，将收到的数据显示在文本框中。

【例 10.1.1】 建立客户端程序实例

```java
package com.example.socketclient;
import android.os.Bundle;
import android.support.v7.app.AppCompatActivity;
import android.widget.TextView;
import java.io.BufferedReader;
import java.io.IOException;
import java.io.InputStreamReader;
import java.net.Socket;
import java.net.SocketTimeoutException;
public class ClientActivity extends AppCompatActivity {
    TextView tvInfo;
    @Override
    protected void onCreate(Bundle savedInstanceState) {
        super.onCreate(savedInstanceState);
        setContentView(R.layout.activity_client);
        tvInfo = (TextView)findViewById(R.id.receivedInfo);
        new Thread() {
            public void run() {
                try{
                    // 创建 Socket
                    Socket socket = new Socket("192.168.252.1", 4666);
                    socket.setSoTimeout(3000);        // 将超时时间设置为 3s
                    BufferedReader br = new BufferedReader(new
                            InputStreamReader(socket.getInputStream()));
                    String lien = br.readLine();    // 读取收到的数据
                    tvInfo.setText("Client received: " + lien);
                    br.close();
                    socket.close();
                }
                catch (SocketTimeoutException e){
                    e.printStackTrace();            // 抛出连接超时异常
                }
                catch(IOException e){
                    e.printStackTrace();
                }
            }
```

```
        }.start();
    }
}
```

注意：上面的代码为 Socket 通信开启了一个新的线程，这是由于在进行网络通信这类不稳定且耗时的任务时，可能会对 UI 线程造成阻塞，因此 Android 2.3 之后的平台都不允许在 UI 线程中建立网络连接及数据通信。

可以看到，此时本地主机的 IP 地址配置为"192.168.252.1"，读者可以根据自己的局域网设置进行配置，并且由于使用了网络连接，因此需要注意在 AndroidManifest.xml 文件中添加网络访问权限，代码如下：

```
<uses-permission android:name="android.permission.INTERNET"/>
```

然后编写服务端的代码，可以直接使用 Java 编写服务端的代码，然后在 PC 上运行。在发送数据流时，为了防止乱码，应注意采用 UTF-8 字符集，代码如下所示。

【例 10.1.2】 建立服务端程序实例

```
package com.example.socketserver;
import java.io.IOException;
import java.io.OutputStream;
import java.net.ServerSocket;
import java.net.Socket;
public class Server {
    public static void main(String[] args) throws IOException{
        // TODO Auto-generated method stub
        ServerSocket serverSocket = new ServerSocket(4666);// 指定端口
        while(true){
            Socket socket = serverSocket.accept(); // 获取 Socket 连接请求
            OutputStream os = socket.getOutputStream();      // 获取输出流
            os.write("This is Server! ".getBytes("UTF-8")); // 发送数据流
            os.close();
            socket.close();
        }
    }
}
```

将上述程序在 PC 端运行后，在 Android 端运行客户端程序，即可看到程序运行结果，如图 10.3 所示。

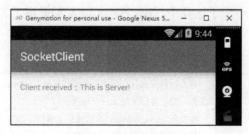

图 10.3 程序运行结果

10.2 使用 URL 访问网络

10.2.1 使用 URL 获取网络资源

URL（Uniform Resource Locator）代表统一资源定位器，它是指向互联网"资源"的指针。资源可以是简单的文件或目录，也可以是对更复杂的对象的引用，如对数据库或搜索引擎的查询。在通常情况下，URL 可以由协议名、主机、端口和资源组成，即满足如下格式：protocol://host:port/resourceName。例如，以下的 URL 地址：

```
http://www.cqupt.com/index.php
```

URL 类提供了多个用于创建 URL 对象的构造方法。一旦获得了 URL 对象之后，就可以调用如表 10.5 所示的常用方法来访问该 URL 对应的资源。

表 10.5 访问 URL 对应资源的常用方法

类 型	方 法	简 述
String	getFile()	获取此 URL 的资源名
String	getHost()	获取此 URL 的主机名
String	getPath()	获取此 URL 的路径部分
String	getProtocol()	获取此 URL 的协议名称
String	getQuery()	获取此 URL 的查询字符串部分
String	getPort()	获取此 URL 的端口号
URLConnection	openConnection()	返回一个 URLConnection 对象，它表示到 URL 所引用的远程对象的连接
InputStream	openStream()	打开与此 URL 的连接，并返回一个用于读取该 URL 资源的 InputStream

在上一节中已经介绍过，在 Android 2.3 之后的平台中，网络访问不能在 UI 线程中进行，所以本节将介绍如何运用 AsyncTask 类创建异步线程并访问 URL 网络资源。

首先了解 AsyncTask 类的基本知识。AsyncTask 类的特点是任务在主线程之外运行，而回调方法是在主线程中执行的，这就有效地避免了使用 Handler 带来的麻烦。AsyncTask 类是抽象类，其中定义了如下 3 种泛型类型。

- Params：启动任务的输入参数，如网络请求的 URL。
- Progress：后台执行任务的百分比，如果不需要，则可以将其定义为 void 类型。
- Result：后台执行任务最终返回的结果，如 String。

AsyncTask 抽象出后台线程运行的 5 个状态，分别是准备运行、正在后台运行、进度更新、完成后台任务、取消任务。对应于这几个状态，AsyncTask 类提供了如表 10.6 所示的常用方法。

表 10.6 AsyncTask 类的常用方法

序 号	方 法	简 述
1	onPreExecute()	该方法将在执行实际的后台操作前被 UI thread 调用。可以在该方法中做一些准备工作，如在界面上显示一个进度条
2	doInBackground(Params...)	在执行 onPreExecute()方法后马上执行，该方法运行在后台线程中，主要负责执行耗时的后台计算工作。可以调用 publishProgress()方法来更新实时的任务进度。该方法是抽象方法，子类必须实现

序号	方法	简述
3	onProgressUpdate(Progress...)	在publishProgress()方法被调用后,UI thread将调用这个方法,从而在界面上展示任务的进展情况,例如,通过一个进度条进行展示
4	onPostExecute(Result)	在doInBackground()方法被执行完成后,onPostExecute()方法将被UI thread调用,后台的计算结果将通过该方法传递到UI thread
5	onCancelled()	取消任务

注意:doInBackground()方法和 onPostExecute()方法的参数必须对应。这些参数在AsyncTask类声明的泛型参数列表中被指定:第一个为doInBackground()方法接收的初始化参数,第二个为显示进度的参数,第三个为doInBackground()方法返回和onPostExecute()方法得到的结果参数。

下面用一个例子来展示异步线程在实际场景中的应用。在这个例子中,将会加载网络中的一张图片,读者可以通过百度浏览器进行图片搜索,右击图片获取图片地址,从而获取该图片的URL资源。

【例10.2】 访问URL网络资源实例

```java
package com.example.url;
import android.graphics.Bitmap;
import android.graphics.BitmapFactory;
import android.os.AsyncTask;
import android.support.v7.app.AppCompatActivity;
import android.os.Bundle;
import android.widget.ImageView;
import java.io.InputStream;
import java.net.MalformedURLException;
import java.net.URL;
public class WebActivity extends AppCompatActivity {
    ImageView imgView;
    Bitmap bm;
    @Override
    protected void onCreate(Bundle savedInstanceState) {
        super.onCreate(savedInstanceState);
        setContentView(R.layout.activity_web);
        imgView = (ImageView)findViewById(R.id.imgView);
        ShowPicture("http://pic.to8to.com/attch/day_160218/201602
            18_d968438a2434b62ba59dH7q5KEzTS6OH.png");
    }
    public void ShowPicture(String url){
        new AsyncTask<String, Void , String>() {
            @Override
            protected String doInBackground(String…params) {
                try{
                    URL url = new URL(params[0]);      // 定义url为第一个参数
                    InputStream is = url.openStream();  // 获取输入流
                    bm = BitmapFactory.decodeStream(is); // 解析图片
                    is.close();                         // 关闭输入流
                }catch (MalformedURLException e){
```

```
                e.printStackTrace();
            }catch (Exception e){
                e.printStackTrace();
            }
            return null;
        }
        @Override
        protected void onPostExecute(String s) {
            super.onPostExecute(s);
            imgView.setImageBitmap(bm);                  // 显示图片
        }
    }.execute(url);
}
```

同样地,也需要在 AndroidManifest.xml 文件中添加网络访问权限,代码如下:

```
<uses-permission android:name="android.permission.INTERNET"/>
```

程序运行结果如图 10.4 所示。

图 10.4 程序运行结果

10.2.2 使用 URLConnection 提交请求

在了解了使用 URL 可实现的一些简单操作之后,可以使用 URL 中引用远程对象的连接 URLConnection,URLConnection 可以向所代表的 URL 发送请求和读取 URL 的资源。在通常情况下,创建一个和 URL 的连接需要如下几个步骤。
- 创建 URL 对象,并通过调用 openConnection()方法获得 URLConnection 对象。
- 设置 URLConnection 对象的参数和普通请求属性。
- 向远程资源发送请求。
- 当远程资源变为可用时,程序可以访问远程资源的头字段并通过输入流来读取远程资源返回的信息。
- 关闭输入流。

注意:在连接到远程资源之前,必须对 URLConnection 进行配置,并且 URLConnection 实例不能被重用,每一个连接都需要使用不同的 URLConnection 实例。

URLConnection 类提供了如表 10.7 所示的常用方法,用于在建立连接之前设置请求头字段。

表 10.7 URLConnection 类的常用方法

类 型	方 法	简 述
InputStream	getInputStream()	获取该 URLConnection 的输入流
OutputStream	getOutputStream()	获取该 URLConnection 的输出流
void	setDoInput(boolean newValue)	设置该 URLConnection 是否允许输入
void	setDoOutput(boolean newValue)	设置该 URLConnection 是否允许输出
void	setReadTimeout(int timeoutMillis)	设置连接期间的最大等待时间
void	setRequestProperty	设置指定的请求头字段的值
void	setUseCaches(boolean newValue)	设置标志以说明该连接是否允许使用缓存

网络请求通常分为两类：GET 请求方式和 POST 请求方式。这两种请求方式的区别如下。

- GET 请求方式：只需要使用 connect()方法建立远程通信，在进行参数传递时只需要显式地将参数添加在 URL 之后，也就是说，在构造 URL 时需要添加参数。示例代码如下：

```
String urlName = url + "?" + param;              // 定义 URL
URL realUrl = new URL(urlName);
URLConnection conn = realUrl.openConnection();   // 打开和 URL 之间的连接
conn.setRequestProperty("accept", "*/*");        // 设置通用的请求属性
…                                                // 设置通用的请求属性
conn.connect();
```

- POST 请求方式：URL 中不需要添加参数，需要在使用 connect()方法之后，获取 URLConnection 对象所对应的输出流来发送网络请求的参数。示例代码如下：

```
URL realUrl = new URL(url);                      // 打开和 URL 之间的连接
URLConnection conn = realUrl.openConnection();
conn.setRequestProperty("accept", "*/*");
…                                                // 设置通用的请求属性
conn.setDoOutput(true);
conn.setDoInput(true);
out = new PrintWriter(conn.getOutputStream());
                                                 // 获取 URLConnection 对象对应的输出流
out.print(param);                                // 发送请求参数
```

POST 请求方式比较常用，原因如下。
- GET 请求方式是从服务器上获取数据的，POST 请求方式是向服务器传送数据的。
- GET 请求方式传送的数据量较小，不能大于 2KB。POST 请求方式传送的数据量较大，一般被默认为不受限制。
- 虽然 GET 请求方式的执行效率比 POST 请求方式高，但是 POST 请求方式的安全性比 GET 请求方式高。

下面用一个例子展示如何运用 URLConnection 提交网络请求。由于该程序需要后台服务器的支持，因此笔者编写了一个 Java Web 应用，然后将笔记本电脑作为服务器，这样才能实现网络请求。下面只是简单地介绍了 Java Web 工程项目的创建，对于如何搭建测试环境，读者可以查阅相关的资料。

1. 创建服务端

读者可以使用 MyEclipse 创建 Java Web 工程项目，用于模拟服务端的 Java Web 应用，将

Java Web 工程项目命名为"MyServer",并修改 Index.jsp 文件,创建两个文本输入框及一个"提交"按钮,作为"登录演示"页面。

【例 10.3】 Java Web 服务端实例

使用 URLConnection 提交请求,服务端代码如下:

```jsp
<%@ page language="java" import="java.util.*" pageEncoding="UTF-8"%>
<html>
  <head>   <title>My JSP 'index.jsp' starting page</title>   </head>
  <body>  登录演示 <br>
      <form action="${pageContext.request.contextPath}
                    /servlet/LoginServlet" method="post">
        账号:<input type = "text" name = "username" ><br>
        密码:<input type = "text" name = "password" ><br>
        <input type = "submit" value = "提交">
      </form>
  </body>
</html>
```

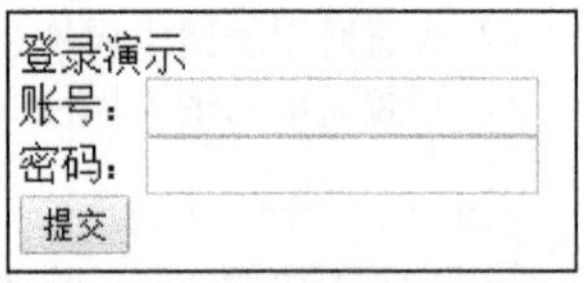

图 10.5 页面显示效果

运行 Tomcat,在浏览器中访问"localhost/Myserver",即可看到如图 10.5 所示的页面显示效果,表示可以成功运行。

然后在该 Java Web 工程项目中,创建一个继承自 HttpServlet 类的 LoginServlet 类,并实现其中的 doPost()方法,用来响应图 10.5 所示页面的用户操作。具体实现代码如下:

```java
package test;
import java.io.IOException;
import java.io.PrintWriter;
import javax.servlet.ServletException;
import javax.servlet.http.HttpServlet;
import javax.servlet.http.HttpServletRequest;
import javax.servlet.http.HttpServletResponse;
public class LoginServlet extends HttpServlet{
public void doPost(HttpServletRequest request, HttpServletResponse response)
            throws ServletException, IOException {
      response.setContentType("text/html;charset=UTF-8");
      request.setCharacterEncoding("UTF-8");
      response.setCharacterEncoding("UTF-8");
      PrintWriter out = response.getWriter();
      String username = request.getParameter("username");  // 获取账号
      String password = request.getParameter("password");  // 获取密码
       // 判断账号、密码是否正确
       if(username.equals("admin") && password.equals("123456")) {
          out.print("Login succeeded!");
       }else {
          out.print("Login failed!");
       }
       out.flush();
       out.close();
    }
}
```

在 Tomcat 中运行上述代码，若在图 10.5 所示的页面中输入"账号"为"admin"、"密码"为"123456"，单击"提交"按钮，则会得到"Login succeeded!"的提示信息，如图 10.6 所示，这样 Java Web 服务端的环境就配置成功了。

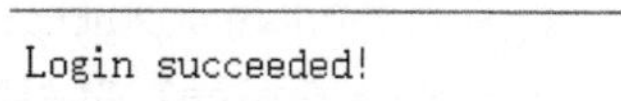

> Login succeeded!

图 10.6　网页登录效果

2. 创建客户端

首先，需要在客户端创建一个登录界面，该界面包含两个文本输入框、一个"提交"按钮和一个返回结果的文本框。然后，在程序中使用异步任务提交网络请求，这里采用的是 POST 请求方式。

【例 10.4.1】 Android 客户端 URLConnection 请求实例

使用 URLConnection 提交请求，代码如下：

```java
package com.example.urlconnection;
import android.os.AsyncTask;
import android.support.v7.app.AppCompatActivity;
import android.os.Bundle;
import android.view.View;
import android.widget.Button;
import android.widget.EditText;
import android.widget.TextView;
import java.io.BufferedReader;
import java.io.IOException;
import java.io.InputStreamReader;
import java.io.PrintWriter;
import java.net.MalformedURLException;
import java.net.URL;
import java.net.URLConnection;
public class URLConnectionDemo extends AppCompatActivity {
    EditText et_username,et_password;
    Button btn_submit;
    TextView tv_result;
    @Override
    protected void onCreate(Bundle savedInstanceState) {
        super.onCreate(savedInstanceState);
        setContentView(R.layout.activity_urlconnection_demo);
        et_username = (EditText)findViewById(R.id.ed_username);
        et_password = (EditText)findViewById(R.id.ed_password);
        btn_submit = (Button)findViewById(R.id.btn_submit);
        tv_result=(TextView)findViewById(R.id.tv_result);
        btn_submit.setOnClickListener(new View.OnClickListener() {
            @Override
            public void onClick(View v) {
                loginCheck(et_username.getText().toString(),
                        et_password.getText().toString());
            }
        });
    }
    public void loginCheck(String username,String password)
```

```java
        {
            final String url =
                "http://172.0.0.1:8888/MyServer/Servlet/LoginServlet"; // 访问的 URL
            final String param = "username=" + username + "&"
                        + "password=" + password;                // 设置参数
            new AsyncTask<String, Void, String>(){
                @Override
                protected String doInBackground(String… params) {
                    PrintWriter out = null;
                    BufferedReader in = null;
                    String result = null;
                    try {
                        URL url = new URL(params[0]);
                        URLConnection conn = url.openConnection();
                        // 设置 URLConnection 参数
                        conn.setRequestProperty("accept","/*");
                        conn.setRequestProperty("connection","Keep-Alive");
                        conn.setRequestProperty("user-agent",
                    "Mozilla/4.0 (compatible; MSIE 6.0; Windows NT 5.1; SV1)");
                        conn.setDoInput(true);     // 允许输入
                        conn.setDoOutput(true);    // 允许输出
                        out = new PrintWriter(conn.getOutputStream());
                        out.print(param);          // 设置参数
                        in = new BufferedReader(new InputStreamReader(conn
                                    .getInputStream()));
                        String line;
                        while((line= in.readLine())!=null)
                        {
                            result += line;        // 获取返回结果
                        }
                    } catch (MalformedURLException e) {
                        e.printStackTrace();
                    } catch (IOException e) {
                        e.printStackTrace();
                    }
                    return result;
                }
                @Override
                protected void onPostExecute(String result) {
                    if(result!=null){
                        tv_result.setText("验证结果：" + result); // 显示返回结果
                    }
                    super.onPostExecute(result);
                }
            }.execute(url);
        }
    }
```

同样地，也需要在 AndroidManifest.xml 文件中添加网络访问权限，代码如下：

```
<uses-permission android:name="android.permission.INTERNET"/>
```

注意：在提交参数时，参数的格式为"param1=value&param2=value…"，中间的"&"用于分隔参数，不能省去。

程序运行结果如图 10.7 所示。

图 10.7　程序运行结果

10.2.3　使用 HttpURLConnection 实现网络通信

在上一节中，展示了如何使用 URLConnection 提交请求。这一节将介绍如何使用 HttpURLConnection 访问网络。HttpURLConnection 类继承自 URLConnection 类，新增了一些用于操作 HTTP 资源的方法，如表 10.8 所示。

表 10.8　HttpURLConnection 类的常用方法

类　型	方　法	简　述
Int	getResponseCode()	获取服务器的响应代码
String	getResponseMessage()	获取服务器的响应消息
String	getRequestMethod()	获取发送请求的方法
void	setRequestMethod(String method)	设置发送请求的方法

下面将采用 HttpURLConnection 类所提供的方法实现之前提到的网络请求，只需要对异步请求的代码进行如下修改。

【例 10.4.2】　Android 客户端 HttpURLConnection 请求实例

使用 HttpURLConnection 提交请求，代码如下：

```java
public void loginCheck2(String username,String password)
{
    final String url = "http://172.0.0.1:8888/MyServer/Servlet/
                        LoginServlet";                    // 访问的 URL
    final String param = "username=" + username + "&"
                        + "password=" + password;          // 设置参数
    new AsyncTask<String, Void, String>(){
        @Override
        protected String doInBackground(String... params) {
            PrintWriter out = null;
            BufferedReader in = null;
            String result = null;
```

```java
            try {
                URL url = new URL(params[0]);
                HttpURLConnection httpConn =
                        (HttpURLConnection) url.openConnection();
                // 设置URLConnection参数
                conn.setRequestProperty("accept","/*");
                conn.setRequestProperty("connection","Keep-Alive");
                conn.setRequestProperty("user-agent",
          "Mozilla/4.0 (compatible; MSIE 6.0; Windows NT 5.1; SV1)");
                httpConn.setDoInput(true);           // 允许输入
                httpConn.setDoOutput(true);          // 允许输出
                httpConn.setRequestMethod("post");   // 设置为POST请求方式
                httpConn.setUseCaches(false);  // 使用POST请求方式时不能使用缓存
                out = new PrintWriter(httpConn.getOutputStream());
                out.print(param);                    // 设置参数
                int responseCode = httpConn.getResponseCode();
                if(responseCode==HttpURLConnection.HTTP_OK){
                    in = new BufferedReader(new
                        InputStreamReader(httpConn.getInputStream()));
                    String line;
                    while((line= in.readLine())!=null)
                    {
                        result += line;              // 获取返回结果
                    }
                }
            }
            catch (MalformedURLException e) {
                e.printStackTrace();
            } catch (IOException e) {
                e.printStackTrace();
            }
            return result;
        }
        @Override
        protected void onPostExecute(String result) {
            if(result!=null){
                tv_result.setText("验证结果:" + result); // 显示返回结果
            }
            super.onPostExecute(result);
        }
    }.execute(url);
}
```

运行上述代码，同样可以得到如图 10.7 所示的结果。有的读者可能在以前学习 Android 网络时接触过 HttpClient 类，但是 Google 在 Android API 23 之后就不在 SDK 中自带 HttpClient 类了，原因是 Apache 不再维护 HttpClient 类了，官方推荐使用的是 HttpURLConnection 类。如果仍然需要使用 HttpClient 类，最便捷的方法是在 build.gradle 中的 android {}中加上"useLibrary "org.apache.http.legacy""，有需要的读者可以自行尝试。

10.3 使用 WebView

随着后台技术的不断发展，App 前端应用都内置了 Web 应用的界面，这个界面就是由 WebView 渲染出来的。这样在开发一些类似于帖子、活动页面的功能时，就可以使用后台统一开发，而对于不同的前端，只需直接调用即可，极大地优化了开发效率。WebView 有如下几个优点：可以直接显示和渲染 Web 应用的界面或网页；可以直接调用网络上或本地的 HTML 文件，也可以和 JavaScript 交互调用。

10.3.1 使用 WebView 浏览网页

WebView 类类似于一个小型的浏览器，提供了一些类似于浏览器的常用方法，如表 10.9 所示。

表 10.9 WebView 类的常用方法

类型	方法	描述
void	goBack()	返回
void	goForward()	前进
void	getProgress()	获取当前访问进度
void	getTitle()	获取当前访问页面的标题
void	loadUrl(String url)	加载指定的页面
void	reload()	重新加载当前页面
void	savePassword(String host, String username, String password)	保存密码
boolean	zoomIn()	放大网页
boolean	zoomOut()	缩小网页
WebSetting	getSettings()	返回 WebSetting 对象

下面使用 WebView 做一个迷你浏览器，在布局界面中放入"后退""前进""刷新""前往"按钮，以及网址输入框，此处不再对布局界面进行赘述，主程序代码如下所示。

【例 10.5】 使用 WebView 加载网页视图实例

```
package com.example.webview;
import android.support.v7.app.AppCompatActivity;
import android.os.Bundle;
import android.view.KeyEvent;
import android.view.View;
import android.webkit.WebSettings;
import android.webkit.WebView;
import android.webkit.WebViewClient;
import android.widget.Button;
import android.widget.EditText;
public class MiniBrowser extends AppCompatActivity implements
                        View.OnClickListener{
    Button btnBack,btnForward,btnRefresh,btnGo;
    EditText etURL;
    private WebView webView;
    @Override
    protected void onCreate(Bundle savedInstanceState) {
        super.onCreate(savedInstanceState);
```

```java
        setContentView(R.layout.activity_mini_browser);
        btnBack = (Button)findViewById(R.id.btnBac);
        btnForward = (Button)findViewById(R.id.btnFow);
        btnRefresh = (Button)findViewById(R.id.btnRef);
        btnGo = (Button)findViewById(R.id.btnGo);
        etURL = (EditText)findViewById(R.id.etUrl);
        webView = (WebView)findViewById(R.id.webView);
        btnBack.setOnClickListener(this);
        btnForward.setOnClickListener(this);
        btnRefresh.setOnClickListener(this);
        btnGo.setOnClickListener(this);
    }
    @Override
    public void onClick(View v) {
        switch (v.getId()){
            case R.id.btnBac:
                webView.goBack();
                break;
            case R.id.btnFow:
                webView.goForward();
                break;
            case R.id.btnRef:
                webView.reload();
                break;
            case R.id.btnGo:
                if(etURL.getText().toString()!=null){
                    // 加载需要显示的网页
                    webView.loadUrl("http://"+etURL.getText().toString());
                    // 设置Web视图
                    webView.setWebViewClient(new webViewClient());
                }
                break;
            default: break;
        }
    }
    @Override
    // 设置回退
    public boolean onKeyDown(int keyCode, KeyEvent event) {
        if ((keyCode == KeyEvent.KEYCODE_BACK) && webView.canGoBack()) {
            webView.goBack(); // goBack()方法表示返回WebView的上一个页面
            return true;
        }
        return false;
    }
    // Web视图
    private class webViewClient extends WebViewClient {
        public boolean shouldOverrideUrlLoading(WebView view, String url) {
            view.loadUrl(url);
            return true;
        }
    }
}
```

程序运行结果如图10.8所示。该程序类似于一个迷你浏览器,在输入网址后,单击"前往"按钮,即可看到下方的 WebView 加载的网页,单击上方的"后退""前进""刷新"按钮可以执行对应的操作。

图 10.8 程序运行结果

10.3.2 加载本地 HTML 网页

在应用的实际开发过程中,某些界面可以直接加载 HTML 网页,如注册说明或使用帮助界面,这样可以在一次开发后将其直接用于多个平台,十分便捷。下面介绍如何在本地工程项目中添加 HTML 文件,并在程序中调用、显示相应的网页。

首先,在工程项目目录下创建 Assets 文件目录,用于存放 HTML 资源,创建方法为:右击工程项目,在弹出的快捷菜单中选择"New"→"Folder"→"Assets Folder"命令,如图 10.9 所示。

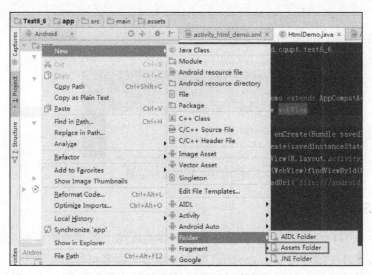

图 10.9 创建 Assets 文件目录

然后，使用文本编辑器编写 HTML 文件，代码如下：

```html
<!DOCTYPE html>
<html>
    <head>
        <meta http-equiv="Content-Type" content="text/html; charset=GBK">
        <title>Insert title here</title>
    </head>
    <body >
        加载本地 HTML 网页！
    </body>
</html>
```

注意：为了防止乱码，字符集选择 GBK 格式。然后在布局界面中插入 WebView 控件，最后定义主程序，代码如下所示。

【例 10.6】 使用 WebView 加载本地 HTML 网页实例

```java
package com.example.webviewlocal;
import android.support.v7.app.AppCompatActivity;
import android.os.Bundle;
import android.webkit.WebView;
public class HtmlDemo extends AppCompatActivity {
    private WebView webView;
    @Override
    protected void onCreate(Bundle savedInstanceState) {
        super.onCreate(savedInstanceState);
        setContentView(R.layout.activity_html_demo);
        webView = (WebView)findViewById(R.id.webview);
        // 加载本地 HTML 网页
        webView.loadUrl("file:///android_asset/index.html");
    }
}
```

程序运行结果如图 10.10 所示。

图 10.10 程序运行结果

10.3.3 JavaScript 交互调用

上述加载本地 HTML 网页的示例只是加载了界面视图，若涉及更多复杂的页面，就无法加载了。HTML 文件中通常带有 JavaScript 脚本。细心的读者可能已经发现了，在加载网页视图时，由于没有对 WebView 进行设置，因此在某些界面会看到如图 10.11 所示的界面。

图 10.11 加载 HTML 网页的提示信息

这时，就需要使用 WebSetting 类对 WebView 进行设置以开启对 JavaScript 的支持，而使用 WebView 类提供的 getSettings()方法可以获取该对象。表 10.10 所示为 WebSetting 类的常用方法。

表 10.10 WebSetting 类的常用方法

类　型	方　　法	简　　述
void	setAllowFileAccess(boolean allow)	设置是否可以访问文件
void	setBuiltInZoomControls(boolean enabled)	设置是否支持缩放操作
void	setBlockNetworkImage(boolean flag)	设置是否显示网络图像
void	setJavaScriptEnabled(boolean enabled)	设置是否启用 JavaScript 支持
void	setDefaultFontSize(int size)	设置默认字号大小
void	setDefaultTextEncodingName(String encoding)	设置默认字符集格式
void	setGeolocationEnabled(boolean enabled)	设置是否可以获取地理位置信息
void	getTitle()	获取当前访问页面的标题

所以，为了开启对 JavaScript 的支持，在例 10.5 中还需要在 loadUrl()方法之前设置 WebView 属性，方法如下：

```
WebSettings webSettings = webview.getSettings();
// 设置WebView属性，能够执行JavaScript脚本
webSettings.setJavaScriptEnabled(true);
// 设置可以访问文件
webSettings.setAllowFileAccess(true);
// 设置支持缩放
webSettings.setBuiltInZoomControls(true);
```

在 HTTP 网页中植入 JavaScript 的最大好处是能和 Android 进行交互操作，如在 App 中弹出 Toast，或者展示列表，或者调用手机的通话功能并直接拨打页面上的电话号码。下面用一个例子展示 JavaScript 与 Android 的交互。

【例 10.7】 JavaScript 与 Android 交互实例

定义 HTML 网页，代码如下：

```
<!DOCTYPE html>
<html>
<head>
    <meta http-equiv="Content-Type" content="text/html; charset=UTF-8">
    <title>Insert title here</title>
```

```html
</head>
    <body >
    点击呼叫电话
    <!--这是一个超链接,调用了Android代码中的call()方法-->
    <a href="javascript:demo.call('10086')">10086</a>
    </body>
</html>
```

Android 端代码如下:

```java
package com.example.javascript;
import android.Manifest;
import android.content.Intent;
import android.content.pm.PackageManager;
import android.net.Uri;
import android.support.v4.app.ActivityCompat;
import android.support.v7.app.AppCompatActivity;
import android.os.Bundle;
import android.webkit.JavascriptInterface;
import android.webkit.WebSettings;
import android.webkit.WebView;
public class WebViewCall extends AppCompatActivity {
    private WebView webView;
    @Override
    protected void onCreate(Bundle savedInstanceState) {
        super.onCreate(savedInstanceState);
        setContentView(R.layout.activity_web_view_call);
        webView = (WebView) findViewById(R.id.webview);
        WebSettings ws = webView.getSettings();
        ws.setJavaScriptEnabled(true);                  // 启动 JavaScript
        ws.setDefaultTextEncodingName("UTF-8");         // 定义字符集为"UTF-8"
        webView.loadUrl("file:///android_asset/call.html");// 加载 HTML 网页
        webView.addJavascriptInterface(this,"demo");// HTML 网页中的 JavaScript
    }
    @JavascriptInterface      // 只有添加这个标签,才能将该方法暴露给 JavaScript
    public void call(final String phone) {
        if (ActivityCompat.checkSelfPermission(this  // 检测权限
                , Manifest.permission.CALL_PHONE) !=
                PackageManager.PERMISSION_GRANTED) {
            return;}
        Intent intent = new Intent(Intent.ACTION_CALL, Uri.parse
                ("tel:" + phone));
        startActivity(intent);
    }
}
```

注意:在上述代码中,在需要暴露给 JavaScript 的方法之前必须加上"@JavascriptInterface"标签,而且需要在 AndroidManifest.xml 文件中添加通话权限,代码如下:

```xml
<uses-permission android:name="android.permission.CALL_PHONE"/>
```

程序运行结果如图 10.12 所示。

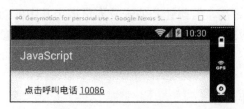

图 10.12　程序运行结果

10.4　使用 WebService 进行网络编程

10.4.1　WebService 基础

前面介绍了多种与后台服务端进行通信的操作，但是在该开发模式下会存在平台的约束。例如，之前开发的 Java Web 应用程序只适用于使用 Java 编写的客户端。如果一个应用程序需要跨平台操作，则可以采用 WebService 作为后台服务端，采用 SOAP（简单对象访问协议）进行通信，通过统一的 XML 语言传输数据。

简单来说，WebService 就是一个网络应用程序，它向外部暴露了能够通过 Web 调用的 API 接口。因为需要实现互操作性，所以 WebService 平台需要一套可创建分布式应用程序的协议。该平台具有以下三大核心技术。

1．XML+XSD

WebService 采用 HTTP 协议传输数据，采用 XML 封装数据。XML 的优点在于易于创建、便于分析，以及具有很好的平台无关性，使任何平台都可以轻松地解析数据。而 XML 没有定义标准的数据类型，正好与 XSD 相结合，解决了数据类型的定义问题。

2．SOAP（简单对象访问协议）

SOAP 是用于交换数据的一种协议规范，是一种轻量级的、基于 XML 的协议，其数据传输可以通过 HTTP 或 SMTP 协议进行。SOAP 消息通常采用从发送端到接收端的单向传输。一条完整的 SOAP 消息包含以下几个元素。

（1）必需的 Envelope 元素，可把此 XML 文档标识为一条 SOAP 消息。
（2）可选的 Header 元素，包含头部信息。
（3）必需的 Body 元素，包含所有的调用和响应信息。
（4）可选的 Fault 元素，提供关于在处理此消息时所发生错误的信息。
其格式如下：

```
<SOAP-ENV:Envelope
  各种属性>
<!--示例-->
  <SOAP:HEADER>
  </SOAP:HEADER>
  <SOAP:Body>
  </SOAP:Body>
</SOAP-ENV:Envelope>
```

3. WSDL（WebService 描述语言）

当完成一个 WebService 后，若想让其他开发人员调用其中的功能，则需要进行一定的规范性操作。基于 XML 的 WSDL 用于描述 WebService 及其函数、参数与返回值。有些软件能够直接获取或生成 WSDL，这是十分方便的。WSDL 描述了 WebService 的 3 个基本属性，如下所述。

（1）服务所提供的操作。
（2）如何访问服务。
（3）服务位于何处（通过 URL 来确定）。

10.4.2 调用 WebService

在 Android 平台调用 WebService 时，需要依赖第三方类库 ksoap2-android。ksoap2-android 是 Google 为 Android 平台开发 WebService 客户端提供的开源项目，所以在开发之前，需要将 ksoap2-android 的 JAR 包添加到 Android 工程项目中，步骤如下。

（1）在 Google 提供的项目下载网站上下载开发包，网址为"https://code.google.com/archive/p/ksoap2-android/source"，选择"Downloads"选项，并选择"ksoap2-android-assembly-2.4-jar-with-dependencies.jar"开发包，如图 10.13 所示。

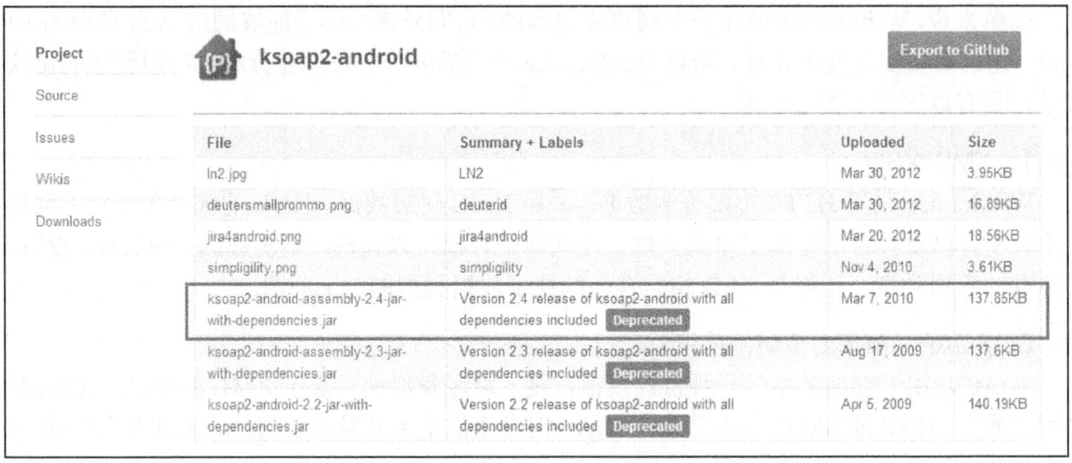

图 10.13　下载 ksoap2-android 的开发包

也可以直接登录华信教育资源网（www.hxedu.com.cn）找到该 JAR 包。

（2）将下载的 ksoap2-android 的 JAR 包添加到工程项目的 libs 目录下，并右击该 JAR 包，在弹出的快捷菜单中选择"Add as library"命令，这样就将 ksoap2-android 集成到 Android 项目中了，如图 10.14 所示。

使用 ksoap2-android 调用 WebSerice 的操作步骤如下。

- 创建 HttpTransportSE 传输对象，传入 WebService 服务器地址，例如：

```
final HttpTransportSE httpSE = new HttpTransportSE(SERVER_URL);
```

- 创建 SoapObject 对象，并在创建该对象时传入所要调用的 WebService 命名空间、WebService 方法名，例如：

```
SoapObject soapObject = new SoapObject(PACE, M_NAME);
```

第 10 章 Android 网络编程基础

图 10.14 将 ksoap2-android 集成到 Android 项目中

- 调用 SoapObject 对象的 addProperty(String name,Object value)方法设置参数。该方法的 name 为参数名；value 为参数值。例如：

```
soapObject.addProperty("byProvinceName", citys);
```

- 创建 SoapSerializationEnelope 对象，并传入 SOAP 协议的版本号；设置对象的 bodyOut 属性，例如：

```
final SoapSerializationEnvelope soapserial = new
            SoapSerializationEnvelope(SoapEnvelope.VER11);
soapserial.bodyOut = soapObject;
soapserial.dotNet = true;
```

- 调用 HttpTransportSE 对象的 call()方法（该方法的第一个参数为 soapAction，第二个参数为 SoapSerializationEvelope 对象），并调用远程 WebService，例如：

```
httpSE.call(PACE + M_NAME, soapserial);
```

- 在调用之后，可以访问 SoapSerializationEnvelope 对象的 bodyIn 属性。该属性将返回一个 SoapObject 对象，我们将其解析后即可获得结果。例如：

```
SoapObject result = (SoapObject) soapserial.bodyIn;
SoapObject detail = (SoapObject) result.getProperty
            ("getSupportProvinceResult");
for (int i = 0; i < detail.getPropertyCount(); i++) {
    citys.add(detail.getProperty(i).toString());
}
```

10.4.3 实现手机归属地查询

在了解了 WebService 的基本访问方法之后，就可以根据网上开放的 WebService 服务开发自己的应用程序了。本节将展示如何实现一个手机归属地查询的小应用，使用的 WebService 服务地址为：http://ws.webxml.com.cn/WebServices/MobileCodeWS.asmx。

在打开该网址后，单击 "getMobileCodeInfo" 链接，进入 WebService 服务页，如图 10.15 所示。

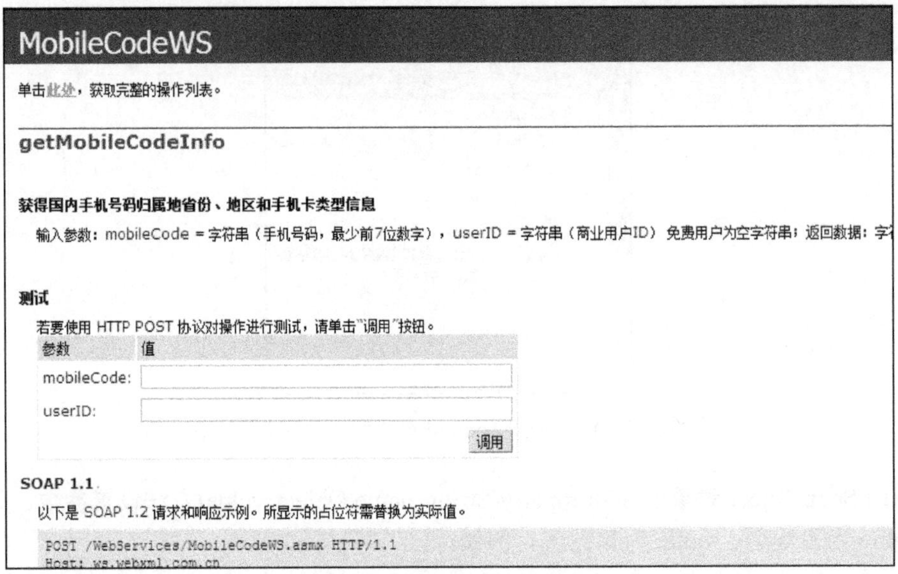

图 10.15　WebService 服务页

在 http://ws.webxml.com.cn/WebServices/MobileCodeWS.asmx 后面加上"?wsdl"，即可访问其 wsdl 说明页，如图 10.16 所示。

图 10.16　wsdl 说明页

从图 10.16 中可以得到以下几个很关键的点。
- 作用域 targetNamespace = http://WebXml.com.cn/。
- 查询的方法名为"getMobileCodeInfo"，需要带上"mobileCode"与"userID"两个参数。
- 返回的结果被存放在"getMobileCodeInfoResult"中。

【例 10.8】　使用 WebService 实现手机归属地查询实例

代码如下：

```
package com.example.webservice;
import android.os.AsyncTask;
import android.support.v7.app.AppCompatActivity;
import android.os.Bundle;
import android.view.View;
import android.widget.Button;
```

```java
import android.widget.EditText;
import android.widget.TextView;
import org.ksoap2.SoapEnvelope;
import org.ksoap2.SoapFault;
import org.ksoap2.serialization.SoapObject;
import org.ksoap2.serialization.SoapSerializationEnvelope;
import org.ksoap2.transport.HttpTransportSE;
import org.xmlpull.v1.XmlPullParserException;
import java.io.IOException;

public class WebClient extends AppCompatActivity {
    // URL 地址
    private static final String SERVER_URL =
        "http://ws.webxml.com.cn/WebServices/MobileCodeWS.asmx?wsdl";
    // 调用的 WebService 命名空间
    private static final String PACE = "http://WebXml.com.cn/";
    // 获取归属地的方法名
    private static final String W_NAME = "getMobileCodeInfo";
    private EditText etPhone;
    private Button btnSearch;
    private TextView tvInfo;
    @Override
    protected void onCreate(Bundle savedInstanceState) {
        super.onCreate(savedInstanceState);
        setContentView(R.layout.activity_web_client);
        etPhone = (EditText) findViewById(R.id.etphone);
        btnSearch = (Button) findViewById(R.id.btnsearch);
        tvInfo = (TextView) findViewById(R.id.tvinfo);
        btnSearch.setOnClickListener(new View.OnClickListener() {
            @Override
            public void onClick(View v) {
                String cityName = etPhone.getText().toString();
                if (cityName.length() > 0) {
                    getWeatherInfo(etPhone.getText().toString());
                }
            }
        });
    }
    private void getWeatherInfo(String phoneMum){
        new AsyncTask<String, Void, String>() {
            @Override
            protected String doInBackground(String... params) {
                String local = "";
                final HttpTransportSE httpSe = new
                            HttpTransportSE(SERVER_URL);
                httpSe.debug = true;
                SoapObject soapObject = new SoapObject(PACE, W_NAME);
                // 定义 soapObject 对象
                soapObject.addProperty("mobileCode", params[0]); // 输入参数
```

```
            soapObject.addProperty("userID", "");
            // 版本号为10
            final SoapSerializationEnvelope serializa = new
                    SoapSerializationEnvelope(SoapEnvelope.VER10);
            serializa.setOutputSoapObject(soapObject);
            serializa.dotNet = true;
            // 获取返回信息
            try {
                httpSe.call(PACE + W_NAME, serializa);
                if (serializa.getResponse() != null) {
                    SoapObject result = (SoapObject) serializa.bodyIn;
                    local = result.getProperty("getMobileCodeInfoResult")
                                        .toString();
                }
            }
            catch (XmlPullParserException e) {
                e.printStackTrace();
            }catch (SoapFault soapFault) {
                soapFault.printStackTrace();
            }catch (IOException e) {
                e.printStackTrace();
            }
            return local;
        }
        @Override
        protected void onPostExecute(String result) {
            tvInfo.setText(result);
        }
    }.execute(phoneMum);
    }
}
```

程序运行结果如图 10.17 所示。

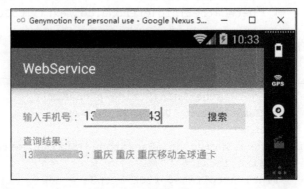

图 10.17　程序运行结果

第 11 章 Android 数据存储

11.1 使用 SharedPreferences

在 Android 中,很多应用都需要存储一些参数,例如,用户在一次使用天气 App 时添加了多个城市,当用户下一次打开该 App 时也希望之前设置的城市会保留在手机中以方便直接获取信息。这时就需要 SharedPreferences 类的辅助,用户可以利用它来存储一些键-值(key-value)对参数。SharedPreferences 类是 Android 提供的一个轻量级的存储类,特别适合用于存储软件的各项参数。

11.1.1 SharedPreferences 的使用方法

首先,获取 SharedPreferences 实例对象。只需要通过 getSharedPreferences(String, int)方法就可以直接获取其实例对象。该方法中的第一个参数用于指定该参数的名称,名称不用带后缀,后缀会由 Android 自动加上;第二个参数用于指定文件的操作模式,共有 4 种操作模式。
- MODE_APPEND:追加方式存储。判断是否有该参数,如果有,则在后面添加。
- MODE_PRIVATE:私有方式存储。其他应用无法访问。
- MODE_WORLD_READABLE:表示当前文件可以被其他应用读取。
- MODE_WORLD_WRITEABLE:表示当前文件可以被其他应用写入。

其次,设置参数。在设置参数时,必须通过一个 SharedPreferences.Editor 对象来设置,以确保在存取时保持参数一致,其操作方法如下:

```
Editor editor = sharedPreferences.edit();
```

Editor 对象采用键-值(key-value)对存放参数,例如,editor.putString("name", "hello")表示只对字符串类型的参数进行存放,当然还可以存放更多的类型,方式都为 editor.putXxx("key","value")的形式,Xxx 可以为 Boolean、Float、Int、Long、String。

在使用 commit()方法提交数据后,可以使用 clean()与 remove()方法清除设置的参数。

例如,通过下面的代码就可以在 Android 的/data/data/<package name>/shared_prefs 目录下生成一个 XML 文件:

```
SharedPreferences sharedPreferences = getSharedPreferences("preference1",
        Context.MODE_PRIVATE);                    // 私有数据
Editor editor = sharedPreferences.edit();         // 获取编辑器
editor.putString("name", "preference1");
editor.putInt("age", 10);
editor.commit();                                  // 提交修改
```

生成的 preference1.xml 文件内容如下:

```
<?xml version='1.0' encoding='utf-8' standalone='yes' ?>
<map>
```

```xml
<string name="name">preference1</string>
<int name="age" value="10" />
</map>
```

与存放数据相对应的获取数据的方法为 getXxx("key","value")，其中第二个参数为默认值。如果 preference1 中不存在与之对应的 key，则将返回该默认值。例如，获取上面的参数，代码如下：

```java
SharedPreferences share=getSharedPreferences("preference1",
        Activity.MODE_WORLD_READABLE);
int i=share.getInt("age",0);
```

11.1.2 SharedPreferences 的应用

下面用一个例子说明 SharedPreferences 在记录参数方面的应用。设置用户姓名与年龄并存储它们，在之后打开 App 时就会显示存储的数据。

【例 11.1】 使用 SharedPreferences 存储用户信息实例

```java
package com.example.sharedpreference;
import android.app.Activity;
import android.content.SharedPreferences;
import android.os.Bundle;
import android.view.View;
import android.widget.Button;
import android.widget.EditText;
import android.widget.TextView;
public class MainActivity extends Activity {
    EditText edName,edAge;
    TextView tvInfo;
    Button btnCommit;
    private static final String FILENAME = "info";  // 定义文件名
    @Override
    protected void onCreate(Bundle savedInstanceState) {
        super.onCreate(savedInstanceState);
        setContentView(R.layout.activity_main);
        edName = (EditText)findViewById(R.id.edname);
        edAge = (EditText)findViewById(R.id.edage);
        btnCommit = (Button)findViewById(R.id.btncommit);
        tvInfo = (TextView)findViewById(R.id.tvinfo);
        SharedPreferences preferences = super.getSharedPreferences
            (FILENAME, Activity.MODE_PRIVATE);           // 获取文件
        // 初始化内容
        tvInfo.setText("用户信息 姓名：" + preferences.getString("name",
         "未定义") + ", 年龄：" + preferences.getInt("age", 0) + "。");
        btnCommit.setOnClickListener(new View.OnClickListener() {
            @Override
            public void onClick(View v) {
                SharedPreferences preferences = getSharedPreferences
                    (FILENAME, Activity.MODE_PRIVATE); // 获取文件
                SharedPreferences.Editor edit = preferences.edit();
```

```
                edit.putString("name", edName.getText().toString());
                edit.putInt("age", Integer.parseInt(edAge.getText().toString()));
                edit.commit();        // 提交修改
                tvInfo.setText("用户信息 姓名："+edName.getText().toString()
        +", 年龄: "+edAge.getText().toString());
            }
        });
    }
}
```

在设置"姓名"为"admin"、"年龄"为"12"之后，关闭程序，然后再次打开程序，运行效果如图 11.1 所示。

info.xml 文件被存放在哪里呢？在 Android Studio 中打开 Android Device Monitor，如图 11.2 所示。

图 11.1　程序运行效果

图 11.2　打开 Android Device Monitor

然后选择"File Explorer"命令，进入 data\data\<packagename>\shared_prefs 目录中，就可以看到 info.xml 文件，如图 11.3 所示。

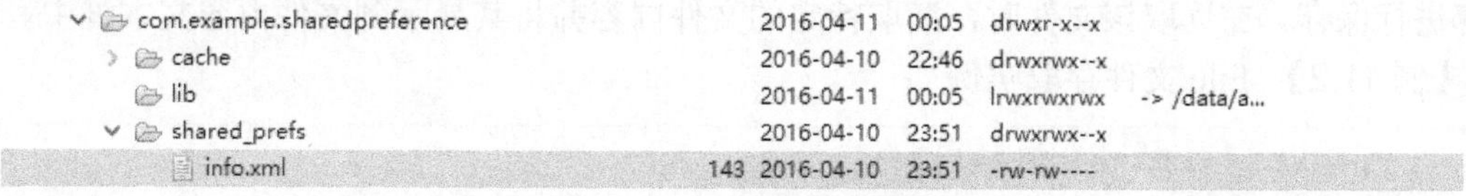

图 11.3　info.xml 文件存储路径

选中该文件，单击窗口右上角的"导出文件"图标就可以将文件导出到计算机上来查看，如图 11.4 所示。

在计算机上查看 info.xml 文件，内容如下：

图 11.4　导出文件

```
<?xml version="1.0" encoding="UTF-8" standalone="true"?>
-<map>
<int value="12" name="age"/>
<string name="name">tt</string>
</map>
```

11.2　File 存储

11.2.1　使用 I/O 流操作文件

Java 中提供的 I/O 流操作在 Android 中也同样适用，使用 Context 所提供的 openFileOutput

与 openFileInput 可以十分方便地操作文件,具体实现过程如下所述。

(1) 文件的存储示例代码如下:

```
// 以 mode 模式获得文件输出流
FileOutputStream out = context.openFileOutput(String filename,int mode);
out.write(byte[]);                                    // 写入内容
```

(2) 文件的读取示例代码如下:

```
// 获得某个文件的文件流
FileInputStream in = context.openFileInput(String filename);
int length = in.read(byte[]);                         // 读取内容
```

(3) 文件的操作模式与之前提到的 SharedPreferences 类中的模式相似,也有 4 种操作模式。

- Context.MODE_PRIVATE:私有覆盖模式。只能被当前应用访问,并且如果写入内容,则会覆盖原有内容。
- Context.MODE_APPEND:私有追加模式。只能被当前应用访问,并且如果写入内容,则会追加该内容到文件中。
- Context.MODE_WORLD_READABLE:公有只读模式。可以被其他应用读取。
- Context.MODE_WORLD_WRITEABLE:公有可写模式。可以被其他应用写入,但不能被其他应用读取。

11.2.2 文件操作应用

下面用一个例子说明 I/O 流在文件操作中的应用。文件将以特定的文件名称和特定的文件内容进行保存。在读取该文件时,读取特定的文件内容并将其显示到文件内容文本框中。

【例 11.2】 File 文件存取实例

```java
package com.example.file;
import android.app.Activity;
import android.content.Context;
import android.os.Bundle;
import android.view.View;
import android.widget.Button;
import android.widget.EditText;
import java.io.ByteArrayOutputStream;
import java.io.FileInputStream;
import java.io.FileOutputStream;
public class MainActivity extends Activity {
    private Button btnSave,btnRead;
    private EditText edFilename,edFilecontent;
    private Context context = this;
    @Override
    public void onCreate(Bundle savedInstanceState) {
        super.onCreate(savedInstanceState);
        setContentView(R.layout.activity_main);
        btnSave = (Button)this.findViewById(R.id.btnSave);
        btnRead = (Button)this.findViewById(R.id.btnRead);
```

```java
edFilename = (EditText)this.findViewById(R.id.edFilename);
edFilecontent = (EditText)this.findViewById(R.id.edFilecontent);
btnSave.setOnClickListener(new View.OnClickListener(){
    @Override
    public void onClick(View v) {
        String filename = edFilename.getText().toString();
        String filecontent = edFilecontent.getText().toString();
        FileOutputStream out = null;
        try {
            out = context.openFileOutput(filename, Context
                    .MODE_PRIVATE);
            out.write(filecontent.getBytes("UTF-8"));
        } catch (Exception e) {
            e.printStackTrace();
        }
        finally{
            try {
                out.close();
            } catch (Exception e) {
                e.printStackTrace();
            }
        }
    }});

btnRead.setOnClickListener(new View.OnClickListener(){
    @Override
    public void onClick(View v) {
        // 获得文件的名称
        String filename = edFilename.getText().toString();
        FileInputStream in = null;
        ByteArrayOutputStream bout = null;
        byte[]buf = new byte[1024];
        bout = new ByteArrayOutputStream();
        int length = 0;
        try {
            in = context.openFileInput(filename);  // 获得输入流
            while((length=in.read(buf))!=-1){
                bout.write(buf,0,length);
            }
            byte[] content = bout.toByteArray();
            // 设置文本框为读取的内容
            edFilecontent.setText(new String(content,"UTF-8"));
        } catch (Exception e) {
            e.printStackTrace();
        }
        edFilecontent.invalidate();                        // 刷新屏幕
        try{
            in.close();
            bout.close();
        }
```

```
            catch(Exception e){}
        }});
   }}
```

对布局文件不再进行赘述，读者可以自行编写，实现效果如图 11.5 所示。

文件存储位置如图 11.6 所示。

导出后即可查看文件内容，如图 11.7 所示。

图 11.6　文件存储位置

图 11.5　实现效果

图 11.7　文件内容

11.2.3　将文件保存到 SD 卡

在上述操作中，直接将文件保存在了默认的存储空间中。如果一个文件很大，则通常被存放在手机的 SD 卡中；如果手机存在 sdcard 目录，则 sdcard 目录为/mnt/sdcard。

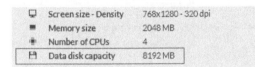

图 11.8　指定存储空间的大小

在 Genymotion 中下载并安装虚拟机的同时，可指定存储空间的大小，如图 11.8 所示。

在 Android 中读取 SD 卡的内容时，需要在 AndroidManifest.xml 文件中设置相关权限，代码如下：

```
<uses-permission android:name=
"android.permission.MOUNT_UNMOUNT_FILESYSTEMS"></uses-permission>
<uses-permission android:name=
"android.permission.WRITE_EXTERNAL_STORAGE"></uses-permission>
```

在 Android 1.5、Android 1.6 中，sdcard 目录为/sdcard，而在 Android 2.0 及以上版本中，sdcard 目录都是/mnt/sdcard。因此，如果在保存时直接写具体目录是不行的，可以使用 Environment 类获取外部存储目录，例如，这个类提供的 getExternalStorageDirectory()方法可以用于直接获取 sdcard 目录。还有一种情况是不确定一部手机是否存在 sdcard 目录，这时也需要使用 Environment 类进行判断。如果这部手机不存在 sdcard 目录，则需要提供其他解决方法。例如，保存到手机存储时提示不存在 sdcard 目录，解决方法如下：

```
if(Environment.getExternalStorageState()
    .equals(Environment.MEDIA_MOUNTED)){
        // 执行存储 sdcard 方法    }
else{   // 存储到手机中，或提示无 SD 卡    }
```

除上述方法外，Environment 类还提供了一些其他方法，可以让用户更方便地操作外部存储目录。Environment 类的常量与常用方法如表 11.1 所示。

表 11.1 Environment 类的常量与常用方法

返回类型	常量或方法名	描述
	MEDIA_BAD_REMOVAL	存储设备被移除
	MEDIA_CHECKING	存储设备正在检查
	MEDIA_MOUNTED	存储设备可读/写
	MEDIA_MOUNTED_READ_ONLY	存储设备只读
	MEDIA_REMOVED	存储设备不存在
static File	getDataDirectory()	获取用户 data 目录
static File	getDownloadCacheDirectory()	获取下载内容目录
static File	getExternalStorageDirectory()	获取扩展存储目录
static String	getExternalStorageState(File path)	获取扩展存储状态
static File	getRootDirectory()	获取根目录

接着上面的例子，可以再添加两个按钮，分别用于将数据保存到 SD 卡中，以及从 SD 卡中读取存储数据。

将数据保存到 SD 卡中，代码如下：

```java
btnSaveSD.setOnClickListener(new View.OnClickListener(){
    @Override
    public void onClick(View v) {
        String filename = edFilename.getText().toString();
        String filecontent = edFilecontent.getText().toString();
        if(Environment.getExternalStorageState().equals(Environment.MEDIA_MOUNTED)){
            File file = new File(Environment.getExternalStorageDirectory().toString()
                + File.separator+"cqupt"+File.separator + filename);  // 定义存储路径
            if(!file.getParentFile().exists()){
                file.getParentFile().mkdirs();                         // 创建目录
            }
            PrintStream out = null;
            try{
                out = new PrintStream(new FileOutputStream(file,true));
                out.println(filecontent);                              // 写入内容
            }catch (Exception e){
                e.printStackTrace();
            }
            finally {
                if (out!= null){
                    out.close();
                }
            }
        }
        else {
            Toast.makeText(MainActivity.this, "SD卡不存在，存储失败",
                Toast.LENGTH_SHORT).show();
        }
    }});
```

从 SD 卡中读取存储数据，代码如下：

```java
btnReadSD.setOnClickListener(new View.OnClickListener() {
    @Override
    public void onClick(View v) {
        // 获得读取的文件的名称
        String filename = edFilename.getText().toString();
        String filecontent = "";
    if(Environment.getExternalStorageState().equals(Environment
            .MEDIA_MOUNTED)){
    File file = new File(Environment.getExternalStorageDirectory().toString()
        + File.separator + "cqupt" + File.separator + filename);
        if (!file.getParentFile().exists()){
            file.getParentFile().mkdirs();
        }
        Scanner scan = null;
        try{
            scan = new Scanner(new FileInputStream(file));
            while(scan.hasNext()){
                filecontent += scan.next()+"\n";
            }
            edFilecontent.setText(filecontent);
        }catch (Exception e){
            e.printStackTrace();
        }finally {
            if (scan!=null){
                scan.close();
            }
        }
    }else {
        Toast.makeText(MainActivity.this, "SD卡不存在，存储失败",
                Toast.LENGTH_SHORT).show();
    }
    }
});
```

上述代码同样实现了文件的读取操作，只不过把数据存放在了外部存储中。

11.3　SQLite 数据库

通过学习上述内容，读者应当已经能够存储简单的键-值对或其他类型的文件了，但在软件项目中，这些是远远不够的。有时我们需要对数据表进行处理，因为一些数据表甚至达到成百上千行，如果没有一个稳健的数据库存储机制，则无法进行这样的存储工作。所以，Android 也支持 SQLite 数据库，用于处理复杂的数据表。SQLite 是轻量级数据存储工具，是遵守 ACID 的关系型数据库管理系统。它具备 SQL 数据库的所有优点，能够通过 SQL 语句操作数据表。

11.3.1 SQLite 数据库介绍

在 Android 上使用 SQLite 之前，需要了解 SQLite 的存储机制及 SQL 常用操作语句。SQLite 是一个开源的嵌入式关系型数据库，是一个可实现自包容、零配置、支持事务的 SQL 数据库引擎。SQLite 在易用性、便携性、紧凑性、高效性和可靠性方面的表现十分优秀。SQLite 的安装和运行非常简单，在大多数情况下只要确保 SQLite 的二进制文件存在即可开始创建、连接和使用数据库。如果需要使用一个嵌入式数据库项目或解决方案，则 SQLite 是绝对值得考虑的。

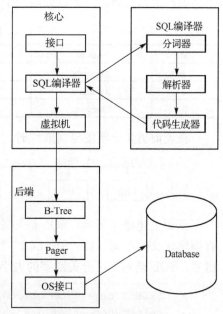

图 11.9 SQLite 体系结构

SQLite 具有十分简洁明了的体系结构，如图 11.9 所示。这些模块的分工明确，将查询过程划分为几个独立的任务，就像在流水线上操作一样。首先，开发人员通过开放的接口与 SQLite 进行交互；其次，经过解析器进行编译，转化为更方便底层处理的层次化数据结构；再次，交由虚拟机完成具体的数据库操作（比如，打开一个表的游标、查询一列等）；最后，后端负责数据的排序、读/写等操作。

在了解 SQLite 的机制之后，还需要学习 SQLite 中的 SQL。SQL 是学习数据库的基础，也是与关系型数据库进行通信的唯一渠道。使用 SQL 能够十分便捷地进行构建、读/写、排序、过滤、计算、分组等常用的信息管理。本节主要介绍 SQLite 中 SQL 的使用，但不进行深入讲解。

1. 语法

SQL 大体上可以分为两个部分：数据操作语言（DML）和数据定义语言（DDL）。

查询和更新指令构成了 DML 部分。

- select：从数据表中获取数据。
- update：更新数据表中的数据。
- delete：从数据表中删除数据。
- insert into：向数据表中插入数据。

SQL 的 DDL 部分用于创建或删除数据表，也可以用于定义索引（键），规定表之间的链接，以及施加表间的约束。

SQL 中比较重要的 DDL 语句如下。

- create database：创建新数据库。
- alter database：修改数据库。
- create table：创建新数据表。
- alter table：变更（改变）数据表。
- drop table：删除数据表。
- create index：创建索引（搜索键）。
- drop index：删除索引。

SQL 采用声明式的语法规则，就像自然语言的表达方式一样，可以清楚地描述需求，而不需要指定如何做。下面以处理一个"grades"表为例，如表 11.2 所示。

表 11.2 "grades"表

id	name	sex	chinese	math
1	Xiaoming	male	89	99
2	Xiaoming	female	78	89
3	Zhangwei	male	85	92

在查询男生小明的语文成绩时，使用 SQL 语句的表达方式如下：

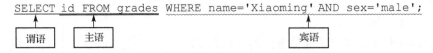

正如上述语句所示，SQL 语句非常简单，其交互方式十分便捷。但要注意，上述语句中有指定含义的单词（如"SELECT""UPDATE""DELETE""INSERT""DROP"等）都是关键字，SQL 语句不区分关键字的大小写，例如，下面的写法表示相同的语句：

```
SELECT name from grades;
seLeCt name From grades;
```

2. 创建数据表

根据上述内容可知，创建数据表的语句属于 DDL 部分。可以使用"CREATE TABLE"创建数据表，并且在创建数据表时需要指定表名、字段名、字段类型及其约束。在 SQLite 中有以下几种数据类型。

- varchar(n)：长度不固定且最大长度为 n 的字串，n 不能超过 4000。
- char(n)：长度固定为 n 的字串，n 不能超过 254。
- integer：值被标识为整数。
- real：所有值都是浮动数值，被存储为 8 字节的 IEEE 浮动标记序号。
- text：值为文本字符串，使用数据库编码进行存储。
- blob：值是 blob 数据块，以输入的数据格式进行存储。如何输入就如何存储，不用改变格式。
- data：包含年份、月份、日期。
- time：包含小时、分钟、秒。

这几种数据类型对于学过数据库的读者来说并不陌生。"约束"用来控制什么样的值可以存储在该字段中。如果需要创建如表 11.2 所示的数据表，则其 SQL 语句如下：

```
create table grades(
    id       integer   identity(1,1) primary key,
    name     varchar   not           null,
    sex      varchar   not           null,
    chinese  integer   not           null,
    math     integer   not           null
)
```

在上述语句中，列属性"identity(起始值,递增量)"表示"id"列为自动编号，因此也将该列称为标识列。

3．数据库常用操作

- 查询：select 作为查询数据库的唯一命令，是 SQL 中功能最强大的但也是最复杂的命令。select 命令采用一系列的子句将很多关系操作组合在一起，从而可以在数据表中只获取所需要的数据。

select 命令的一般格式如下：

```
select [distinct] heading
    from tables                // 指定在哪一个表中
    where predicate            // 对字段进行过滤
    group by columns           // 分组
    having predicate
    order by columns           // 按照什么顺序
    limit count,offset;        // 限定与排序
```

- 增加：insert 命令用于向数据表中插入记录。使用 insert 命令可以一次插入一条记录。insert 命令的一般格式如下：

```
insert into table_name (list1,list2…) values (value1,value2…);
```

变量"table_name"表示将数据插入哪一个数据表中，"list1,list2…"表示用逗号分隔的字段名称。这些字段必须是数据表中存在的，并一一对应于"value1,value2…"。

- 更新：update 命令用于更新数据表中的记录。使用 update 命令可以修改一个数据表中的一行或者多行中的一个或多个字段。update 命令的一般格式如下：

```
update 表名称 set 列名称 = 新值 where 列名称 = 某值;
```

- 删除：delete 命令用于删除数据表中的记录。delete 命令的一般格式如下：

```
delete from 表名称 where 列名称 = 值;
```

11.3.2 SQLite 数据库操作

Android 平台为了方便开发者操作 SQLite，封装了创建和使用 SQLite 数据库的 API，从而可以在应用程序中管理自有的数据库。Android 中提供的 SQLite 操作类如表 11.3 所示。

表 11.3 Android 中提供的 SQLite 操作类

类　名	描　述
SQLiteCursor	游标，用于从数据库中返回查询结果
SQLiteDatabase	提供管理 SQLite 数据库的操作方法
SQLiteOpenHelper	辅助类，用于管理数据库的建立和版本
SQLiteQuery	读取结果行

每一个 SQLiteDatabase 类都代表一个数据库对象，提供了操作数据库的一些方法。在 Android 的 SDK 目录下有 sqlite3 工具，可以利用该工具创建数据库、创建数据表和执行一些 SQL 语句。SQLiteDatabase 类的常用方法如表 11.4 所示。

表 11.4 SQLiteDatabase 类的常用方法

返回类型	方法名	描述
static SQLiteDatabase	openDatabase(String path, SQLiteDatabase.CursorFactory factory, int flags)	打开指定路径下的数据库
static SQLiteDatabase	openOrCreateDatabase(String path, SQLiteDatabase.CursorFactory factory)	打开或创建指定路径下的数据库
long	insert(String table,String nullColumnHack,ContentValues values)	插入一行到数据表中
int	delete(String table,String whereClause,String[] whereArgs)	删除数据表的某一行
int	update(String table,ContentValues values,String whereClause,String[] whereArgs)	更新数据表的某一行
Cursor	query(String table,String[] columns,String selection,String[] selectionArgs, String groupBy,String having,String orderBy)	数据表查询
Cursor	rawQuery(String sql, String[] selectionArgs)	使用 SQL 语句进行数据表查询
void	execSQL(String sql)	执行 SQL 语句
void	setVersion(int version)	设置数据库版本
void	close()	关闭数据库

SQLiteDatabase 类本身只是一个数据库的操作类，如果需要对数据库进行操作，还需要得到 SQLiteOpenHelper（数据库操作辅助类）的帮助。这个类主要用于生成一个数据库，并对数据库的版本进行管理。SQLiteOpenHelper 是一个抽象类，通常需要继承它，并实现以下 3 个函数。

- onCreate(SQLiteDatabase)：在第一次生成数据库时会调用这个方法，也就是说，只有在创建数据库时才会调用该方法。当然，也有一些其他的情况，一般使用这个方法生成数据表。
- onUpgrade(SQLiteDatabase, int, int)：当数据库需要升级时，Android 系统会主动调用这个方法。一般使用这个方法删除数据表，并建立新的数据表。当然，是否还需要进行其他的操作完全取决于应用的需求。
- onOpen(SQLiteDatabase)：这是打开数据库时的回调函数，一般在程序中不常使用。

SQLiteOpenHelper 类也提供了一些其他常用的数据库操作方法，如表 11.5 所示。

表 11.5 SQLiteOpenHelper 类的常用方法

返回类型	方法名	描述
String	getDatabaseName()	获取数据库名称
SQLiteDatabase	getReadableDatabase()	以只读方式创建或打开数据库
SQLiteDatabase	getWritableDatabase()	以修改方式创建或打开数据库
void	onOpen(SQLiteDatabase db)	当数据库已经打开时被调用
int	update(String table,ContentValues values,String whereClause,String[] whereArgs)	更新数据表的某一行
void	close()	关闭数据库

在了解了 SQLite 的常用操作方式之后，就可以开始建立数据库并进行实践练习了。下面通过一个实例来展示如何使用 SQLite。

1. 创建数据库

定义 SQLiteOpenHelper 类的子类，在该子类中创建并打开一个指定名称的数据库对象，代码如下：

```java
package com.example.sqlite;
import android.content.Context;
import android.database.sqlite.SQLiteDatabase;
import android.database.sqlite.SQLiteOpenHelper;
public class MyDatabaseHelper extends SQLiteOpenHelper {
    private static final String DATABASE_NAME = "sqlite.db";
    private static final int DATABASE_VERSION = 1;
    private static final String TABLE_NAME = "grades";
// 调用父类构造器
public MySQLiteHelper(Context context, String name, SQLiteDatabase
.CursorFactory factory, int version) {
    super(context, DATABASE_NAME, factory, DATABASE_VERSION);
}
/**
 * 当数据库首次创建时执行该方法，一般将创建数据表等初始化操作放在该方法中
 * 执行，重写 onCreate()方法，调用 execSQL()方法创建数据表
 */
@Override
public void onCreate(SQLiteDatabase db) {
    db.execSQL("create table" + "if not exists" + TABLE_NAME + "("
        + "id integer primary key,"
        + "name varchar,"
        + "sex varchar,"
        + "chinese integer,"
        + "math integer)");
}
// 当打开数据库时传入的版本号与当前的版本号不同时会调用该方法
@Override
public void onUpgrade(SQLiteDatabase db, int oldVersion, int newVersion) {
    db.execSQL("drop table if exists" + TABLE_NAME);
    this.onCreate(db);          // 创建数据表
}
}
```

然后定义 MainActivity.java，调用 MyDatabaseHelper 类完成数据表的创建，代码如下：

```java
package com.example.sqlite;
import android.support.v7.app.AppCompatActivity;
import android.os.Bundle;
public class MainActivity extends AppCompatActivity {
MySQLiteHelper helper;
public static final String TABLE_NAME = "grades";
    @Override
    protected void onCreate(Bundle savedInstanceState) {
        super.onCreate(savedInstanceState);
        setContentView(R.layout.activity_main);
```

```
            helper = new MySQLiteHelper(this);
            helper.getWritableDatabase();
        }
    }
```

使用上述代码轻松地建立了一个数据库,并在数据库中插入了"grades"表。在 Genymotion 模拟器上运行程序之后,我们就可以采用与 11.1 节中查看存储文件相同的方式获取 db 文件,db 文件的存储路径为 data/data/com.example.sqlite/databases/sqlite.db,如图 11.10 所示。

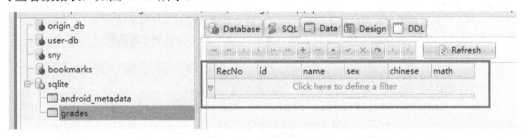

图 11.10 db 文件的存储路径

如果计算机上没有安装相应的软件,则无法打开 db 文件。在安装 SQLite Expert 软件后,即可查看数据表,如图 11.11 所示。

图 11.11 查看数据表

2. 修改数据表中的数据

在创建数据库之后,即可直接使用 SQL 语句操作数据库。在主程序中添加如下数据库修改方法:

```
        public void insert(String name, String sex, int chinese, int math){
            SQLiteDatabase db = helper.getWritableDatabase();
            db.execSQL("insert into "+ TABLE_NAME +"(name,sex,chinese,math)
                values('"+ name +"','" + sex + "','"+ chinese+"','"+ math +"')");
            db.close();
        }

        public void update(int id, String name, String sex, int chinese, int math){
            SQLiteDatabase db = helper.getWritableDatabase();
            db.execSQL("update "+ TABLE_NAME +" set name ='" + name + "',sex='" +
                sex + "',chinese='" + chinese + "',math='" + math +"'where id=" + id);
            db.close();
        }
        public void delete(int id)
        {
            SQLiteDatabase db = helper.getWritableDatabase();
            db.execSQL("delete from " +TABLE_NAME+" where id=" + id);
```

```
        db.close();
    }
```

如果在数据表中依次调用这些函数,则可以看到数据库的内容变化。
执行如下 SQL 语句:

```
insert("Xiaoming","male", 89, 99);
insert("Xiaoming", "female", 78, 89);
insert("Zhangwei","male",85,92);
update(2, "Xiaohong", "female", 85, 97);
delete(3);
```

数据库修改后的结果如图 11.12 所示。

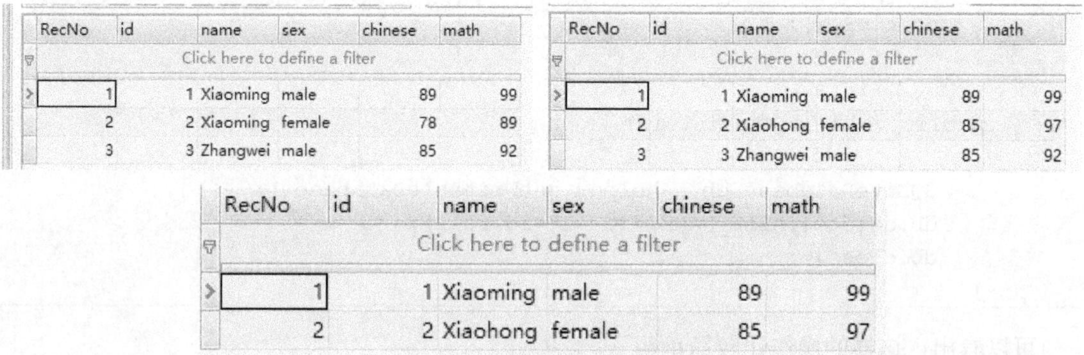

图 11.12 数据库修改后的结果

在上述方法中,直接使用 SQL 语句对数据表进行了操作,可以看到语句中的符号很多,特别容易出错,所以为了更方便地操作数据,SQLiteDatabase 类中也提供了 insert()、update()、delete()等数据操作方法。另外,在这些方法中还可以使用 ContentValues 类封装数据。ContentValues 类的使用方式类似于 SharedPreferences 类中提供的键-值对数据存储。ContentValues 类的常用方法如表 11.6 所示。

表 11.6 ContentValues 类的常用方法

返回类型	方法名	描述
void	clear()	清空数据
boolean	containsKey(String key)	判断是否包含该字段
void	put(String key, Xxx value)	添加值到指定的字段中,Xxx 为字段类型
Xxx	getAsXxx(String key)	获取字段中的值,Xxx 为字段类型
int	size()	返回数值

所以可以对上述的各个方法进行修改,代码如下:

```
public void insert(String name, String sex, int chinese, int math){
    SQLiteDatabase db = helper.getWritableDatabase();
    ContentValues values = new ContentValues();
    values.put("name", name);
    values.put("sex", sex);
    values.put("chinese",chinese);
```

```
        values.put("math",math);
        db.insert(TABLE_NAME, "id", values);
        db.close();
    }
    public void update(int id, String name, String sex, int chinese, int math){
        SQLiteDatabase db = helper.getWritableDatabase();
        ContentValues values = new ContentValues();
        values.put("name", name);
        values.put("sex", sex);
        values.put("chinese",chinese);
        values.put("math",math);
        db.update(TABLE_NAME, values,"id = "+id,null);
        db.close();
    }
    public void delete(int id)
    {
        SQLiteDatabase db = helper.getWritableDatabase();
        db.delete(TABLE_NAME,"id = "+id,null);
        db.close();
    }
```

可以看出,这样的操作比较简单明了。

3. 查询数据表中的数据

在建立好数据表之后,下一步应查询需要的信息。SQL 提供了 select 这个功能强大但复杂的查询语句。与上述操作方式一样,可以使用 SQLiteDatabase 类所提供的数据库操作方法,如 rawQuery()或 Query()方法。在学习这些查询方法之前,需要了解 Cursor 类。Cursor 就是游标,类似于在数据表中游走,用于对结果集进行随机访问,Cursor 类与 JDBC 中 ResultSet 类的作用相似。Cursor 类也提供了一些常用方法,如表 11.7 所示。

表 11.7 Cursor 类的常用方法

返回类型	方法名	描述
void	close()	关闭游标,释放数据库资源
int	getCount()	返回数据表的行数
int	getPosition()	返回游标所在的行数
int	getColumnIndex(String columnName)	返回查询字段的列。当该值不存在时,返回-1
String	getString(int columnIndex)	以字符串的形式返回该列的值
boolean	moveToNext()	移动游标到下一行
boolean	moveToPrevious()	移动游标到上一行
boolean	moveToFirst()	移动游标到第一行
boolean	moveToLast()	移动游标到最后一行

与之前的数据库操作方式一样,既可以使用 SQL 语句,也可以使用 SQLiteDatabase 类所提供的 query()方法进行查询。

假设现在需要查询某一位同学的语文成绩,则可以通过 rawQuery()方法实现带参数的查询，代码如下：

```
public int getChinese(String name){
    int grade = -1;
    Cursor c = db.rawQuery("select * from "+TABLE_NAME+" where name=?",name);
    if(cursor.moveToFirst()) {
        grade = c.getString(c.getColumnIndex("chinese"));   // 获取相关的分数
    }
    return grade;
}
```

或者使用 query()方法，完整带参数的方法如下：

```
query(String table, String[] columns, String selection, String[] selectionArgs, String groupBy, String having, String orderBy, String limit)
```

其中，各个字段的含义如下所述。
- table：数据表名称。
- columns：列名称数组。
- selection：条件子句，相当于 where。
- selectionArgs：条件语句的参数数组。
- groupBy：分组。
- having：分组条件。
- orderBy：排序方式，可以选定一个字段，使其以升序或降序的方式排列。
- limit：分页查询的限制。

上述的查询方法可改为：

```
public int getChinese(){
    // 查询并获得游标
    Cursor c = db.query(TABLE_NAME,null,null,null,null,null,null);
    if(c.moveToFirst()){                // 判断游标是否为空
      grade = c.getString(c.getColumnIndex("chinese"));
    }
}
```

在学会查询语句的使用方式之后，可以完成一个完整的数据库操作小程序。下面完成一个简易的记账软件的开发，在该软件中可以定义金额的数值、金额的类型（支出或收入）和描述，并且可以查询当前的总计支出与总计收入。

【例 11.3】 记账软件实例

首先，创建 DBHelper 类。该类继承于 SQLiteOpenHelper 类，用于创建数据库。DBHelper.java 文件代码如下：

```
package com.example.billbook;
import android.content.Context;
import android.database.sqlite.SQLiteDatabase;
import android.database.sqlite.SQLiteOpenHelper;
public class DBHelper extends SQLiteOpenHelper {
    private static final String DATABASE_NAME = "sqlite.db"; // 数据库名称
```

```java
        private static final String TABLE_NAME = "count";         // 数据表名称
        private static final int VERSION = 1;                      // 数据库版本
        public DBHelper(Context context) {
            super(context, DATABASE_NAME, null, VERSION);
        }
        @Override
        public void onCreate(SQLiteDatabase db) {
            db.execSQL("create table " + "if not exists " + TABLE_NAME + " ("
                    + "id integer primary key,"
                    + "count varchar,"
                    + "type varchar,"
                    + "date varchar,"
                    + "describe varchar)");
        }
        @Override
        public void onUpgrade(SQLiteDatabase db, int oldVersion, int newVersion) {
            db.execSQL("drop table if exists" + TABLE_NAME);
            this.onCreate(db);
        }
    }
```

其次，定义 Count 类，用于封装账单数据。Count.java 文件代码如下：

```java
    package com.example.billbook;
    public class Count {
        private Double money;
        private String type;
        private String date;
        private String describe;
        public Count() {
        }
        public Count(Double money, String date, String type, String describe) {
            this.money = money;
            this.date = date;
            this.type = type;
            this.describe = describe;
        }
        public Double getMoney() { return money; }
        public void setMoney(Double money) { this.money = money; }
        public String getType() { return type; }
        public void setType(String type) { this.type = type; }
        public String getDate() { return date; }
        public void setDate(String date) {this.date = date; }
        public String getDescribe() { return describe; }
        public void setDescribe(String describe) { this.describe = describe; }
    }
```

再次，定义 DBManager 类，用于维护和管理数据库，包括数据插入、数据查询等操作。DBManager.java 文件代码如下：

```java
    package com.example.billbook;
```

```java
import android.content.ContentValues;
import android.content.Context;
import android.database.Cursor;
import android.database.sqlite.SQLiteDatabase;
public class DBManager {
    private DBHelper helper;
    private SQLiteDatabase db;
    private static final String TABLE_NAME = "count";
    public DBManager(Context context) {
        helper = new DBHelper(context);
        db = helper.getWritableDatabase();
    }
    public void insert(Count count) {
        db.beginTransaction();              // 开始事务
        try {
            ContentValues cv = new ContentValues();
            cv.put("count",count.getMoney());
            cv.put("type",count.getType());
            cv.put("date", count.getDate());
            cv.put("describe",count.getDescribe());
            db.insert(TABLE_NAME,"id",cv);
            db.setTransactionSuccessful();   // 设置事务成功完成
        }finally {
            db.endTransaction();             // 结束事务
        }
    }
    public Double getResult(int type)
    {
        Double result = 0.0;
        Cursor c = db.rawQuery("select id,count,type,date
            ,describe from "+ TABLE_NAME,null);
        for (c.moveToFirst();!c.isAfterLast();c.moveToNext()) {
            if (c.getInt(2) == type)
                result += Double.parseDouble(c.getString(1));
        }
        c.close();
        return result;
    }
    public void closeDB(){
        db.close();
    }
}
```

最后，在 MainActivity.java 文件中实例化数据库，并对 UI 控件添加监听功能，完成记账功能。代码如下：

```java
package com.example.billbook;
import android.support.v7.app.AppCompatActivity;
import android.os.Bundle;
import android.view.View;
```

```java
import android.widget.Button;
import android.widget.EditText;
import android.widget.RadioGroup;
import android.widget.TextView;
import java.text.SimpleDateFormat;
import java.util.Date;

public class MainActivity extends AppCompatActivity {
    private static final int OUT = 1;
    private static final int IN = 2;
    private DBManager mgr;
    private EditText etCount,etDescribe;
    private RadioGroup radioGroup;
    private TextView tvInfo;
    private Button btnAdd;
    private int countType = OUT;
    @Override
    protected void onCreate(Bundle savedInstanceState) {
        super.onCreate(savedInstanceState);
        setContentView(R.layout.activity_main);
        // 初始化 DBManager
        mgr = new DBManager(this);
        // 获取 UI 控件
        etCount = (EditText) findViewById(R.id.etCount);
        etDescribe = (EditText) findViewById(R.id.etDescribe);
        radioGroup = (RadioGroup) findViewById(R.id.radioGroup);
        tvInfo = (TextView)findViewById(R.id.tvInfo);
        btnAdd = (Button)findViewById(R.id.btnAdd);
        radioGroup.setOnCheckedChangeListener(new
                RadioGroup.OnCheckedChangeListener() {
            @Override
            public void onCheckedChanged(RadioGroup group, int checkedId) {
                switch (checkedId) {
                    case R.id.radioOut:
                        countType = OUT;// 定义为支出
                        break;
                    case R.id.radioIn:
                        countType = IN; // 定义为收入
                        break;
                }
            }});
        btnAdd.setOnClickListener(new View.OnClickListener() {
            @Override
            public void onClick(View v) {
                Count count = new Count();
                long time = System.currentTimeMillis();    // 获取系统时间
                SimpleDateFormat format = new SimpleDateFormat("yyyy-MM-dd
                                    HH:mm:ss");     // 格式化时间
                String str = format.format(new Date(time)); // 转化为字符串
```

```
                    count.setDate(str);
                    count.setMoney(Double.parseDouble(etCount.getText()
                        .toString()));
                    count.setDescribe(etDescribe.getText().toString());
                    count.setType(countType + "");
                    mgr.insert(count);                          // 插入数值
                    resetInfo();                                // 刷新结果
                }
            });
            resetInfo();
        }
        public void resetInfo(){
            Double out = mgr.getResult(OUT);
            Double in = mgr.getResult(IN);
            Double all = in - out;
            tvInfo.setText("总计支出："+out+"  总计收入："+in+"结余："+all+"。");
        }
        protected void onDestroy() {
            super.onDestroy();
            // 在应用的最后一个 Activity 关闭时应释放 DB
            mgr.closeDB();
        }
    }
```

可以看到，软件显示效果如图 11.13 所示。

图 11.13　软件显示效果

第 12 章 GPS 应用开发

随着智能手机的逐渐普及，为了适应人们的需求，手机定位已经成为智能手机必不可少的功能。GPS 模块集成度高，定位精度高，因此常被用于手机定位。Android 系统提供了基础的 API 以获取经/纬度等数据，这为我们开发应用提供了很大的帮助。

12.1 支持 GPS 的核心 API

类似于前文所述的多媒体支持，Android 提供了一个 AudioManager 服务类。同样地，Android 为了支持 GPS 提供了一个 LocationManager 服务类。所有与 GPS 相关的服务，都需要通过 LocationManager 类来获取。获取该类的方式与获取 AudioManager 类的方式类似，需要调用系统服务。方法如下：

```
LocationManager lm = getService(Context.LOCATION_SERVICE);
```

LocationManager 类提供了丰富的定位服务，其常用方法如表 12.1 所示。

表 12.1　LocationManager 类的常用方法

类 型	方 法	简 述
boolean	addGpsStatusListener(GpsStatus.Listener listener)	添加 GPS 状态监听器
	isProviderEnabled(String provider)	判断指定名称的 LocationProvider 是否可用
void	addProximityAlert(double latitude, double longitude, float radius, long expiration, PendingIntent intent)	添加一个临近警告
	removeGpsStatusListener(GpsStatus.Listener listener)	删除 GPS 状态监听器
	removeProximityAlert(PendingIntent intent)	删除一个临近警告
	requestLocationUpdates(String provider, long minTime, float minDistance, PendingIntent intent)	通过指定的 LocationProvider 周期性地获取定位信息，并通过 intent 启动相应的组件
	requestLocationUpdates(String provider, long minTime, float minDistance, LocationListener listener)	通过指定的 LocationProvider 周期性地获取定位信息，并触发 listener 所对应的触发器
List<String>	getAllProviders()	获取所有的 LocationProvider 列表
	getProviders(Criteria criteria, booleanenabledOnly)	根据指定条件获取满足该条件的全部 LocationProvider 的名称
	getProviders(boolean enabledOnly)	获取所有可用的 LocationProvider
String	getBestProvider(Criteria criteria, boolean enabledOnly)	根据指定条件返回最优的 LocationProvider 对象
GpsStatus	getGpsStatus(GpsStatus status)	获取 GPS 状态
Location	getLastKnownLocation(String provider)	根据 LocationProvider 获取最近一次已知的 Location
LocationProvider	getProvider(String name)	根据名称获取 LocationProvider

可以从表 12.1 中看到一个重要的类——LocationProvider（定位提供者）。这个类提供了获取定位组件信息的功能，以判断手机中的定位组件能提供哪些数据，如能否获取海拔信息等。

它的常用方法如表 12.2 所示。

表 12.2 LocationProvider 类的常用方法

类型	方法	简述
boolean	hasMonetaryCost()	返回该 LocationProvider 是收费的还是免费的
	meetsCriteria(Criteria criteria)	判断该 LocationProvider 是否满足 Criteria 条件
	requiresCell()	判断该 LocationProvider 是否需要访问网络基站
	requiresNetwork()	判断该 LocationProvider 是否需要网络数据
	requiresSatellite()	判断该 LocationProvider 是否需要访问基于卫星的定位系统
	supportsAltitude()	判断该 LocationProvider 是否支持海拔信息
	supportsBearing()	判断该 LocationProvider 是否支持方向信息
	supportsSpeed()	判断该 LocationProvider 是否支持速度信息
int	getAccuracy()	返回该 LocationProvider 的精度
	getPowerRequirement()	获取该 LocationProvider 的电源需求
String	getName()	返回该 LocationProvider 的名称

最后，可以通过 Location 类来获取 GPS 所提供的各项数据，如精确的经/纬度信息、速度信息等。Location 类的常用方法如表 12.3 所示。

表 12.3 Location 类的常用方法

类型	方法	简述
boolean	hasAccuracy()	判断该定位信息是否有精度信息
	hastAltitude ()	判断该定位信息是否有海拔信息
	hasBearing()	判断该定位信息是否有方向信息
	hasSpeed()	判断该定位信息是否有速度信息
double	getAltitude()	获取定位信息的海拔
	getLatitude()	获取定位信息的纬度
	getLongitude()	获取定位信息的经度
float	getAccuracy()	获取定位信息的精度
	getBearing()	获取定位信息的方向
	getSpeed()	获取定位信息的速度
String	getProvider()	获取提供该定位信息的 LocationProvider

在实际应用中，获取 GPS 定位信息的方法主要包括以下几个步骤。
- 获取系统的 GPS 服务类，即 LocationManager 实例对象。
- 获取 LocationProvider，定义指定的定位信息，启动定位服务。
- 使用 Location 获取详细的定位信息。

12.2 获取 LocationProvider

通过上一节的介绍，我们知道 GPS 的定位信息是由 LocationProvider 决定的。这时，如果不进行设置，Android 就会提供默认的定位信息服务。但我们也可以通过设置 LocationProvider 来获取需要的服务。下面就介绍几种获取 LocationProvider 的方法。

1. 获取所有的 LocationProvider

【例 12.1】 获取所有的 LocationProvider 实例

LocationManager 类提供了 getAllProviders()方法，可以用于获取所有的 LocationProvider。下面通过代码说明如何获取所有的 LocationProvider。在主界面中通过一个 ListView 显示了获取的 LocationProvider，布局界面代码十分简单，这里不再展示。代码如下：

```java
package com.example.locationprovider
import android.content.Context;
import android.location.LocationManager;
import android.support.v7.app.AppCompatActivity;
import android.os.Bundle;
import android.widget.ArrayAdapter;
import android.widget.ListView;
import java.util.List;
public class ProviderList extends AppCompatActivity {
    LocationManager lm;
    ListView lvProviderList;
    @Override
    protected void onCreate(Bundle savedInstanceState) {
        super.onCreate(savedInstanceState);
        setContentView(R.layout.activity_provider_list);
        // 获取系统服务
        lm = (LocationManager)getSystemService(Context.LOCATION_SERVICE);
        lvProviderList = (ListView)findViewById(R.id.listViewProvider);
        // 获取 LocationProvider, 保存在 List 中
        List<String> providerList = lm.getAllProviders();
        ArrayAdapter<String> adapter = new ArrayAdapter<String>(this,
         android.R.layout.simple_expandable_list_item_1,providerList);
        lvProviderList.setAdapter(adapter);
    }
}
```

在 Genymotion 模拟器中运行该程序，可以看到模拟器提供了 passive、gps 两种 LocationProvider；在真实手机上运行该程序，可以看到支持的 LocationProvider 又多了一种，即 network，如图 12.1 所示。这是因为手机可以借助移动蜂窝网络或 Wi-Fi 进行基站定位，而这是 Genymotion 模拟器所没有的。

图 12.1 Genymotion 模拟器与手机支持的 LocationProvider

下面说明这几种 LocationProvider 的含义。
- **passive**：由 LocationManager.PASSIVE_PROVIDER 常量表示。
- **gps**：表示通过 GPS 获取定位信息的 LocationProvider 对象，由 LocationManager.GPS_PROVIDER 常量表示。
- **network**：表示通过网络获取定位信息的 LocationProvider 对象，由 LocationManager.NETWORK_PROVIDER 常量表示。

在实际应用中，gps 与 network 是比较常用的。

2. 通过名称获取 LocationProvider

由前面的例子可以看出，使用 getAllProviders()方法可以获取所有的 LocationProvider。我们也可以使用 LocationManager 所提供的 getProvider(String name)方法，指定 LocationProvider 的名称，直接获取需要的 LocationProvider，代码如下：

```
LocationProvider lp = Locationmanager.getProvider
        (LocationManager.GPS_PROVIDER);
```

3. 通过 Criteria 类获取 LocationProvider

使用上述两种方法可以直接获取 LocationProvider，但在实际应用中，我们往往会对定位服务有一定的要求。例如，要求 LocationProvider 能够提供海拔信息，这时就需要使用 Criteria 类，顾名思义就是过滤条件。通过这个类，我们可以选取需要的 LocationProvider。Criteria 类的常用方法如表 12.4 所示。

表 12.4 Criteria 类的常用方法

方法	简述
setAccuracy(int accuracy)	设置对 LocationProvider 的精度要求
setAltitudeRequired(boolean altitudeRequired)	设置对 LocationProvider 能提供的海拔信息的要求
setBearingRequired(boolean bearingRequired)	设置对 LocationProvider 能提供的方向信息的要求
setSpeedRequired(boolean speedRequired)	设置对 LocationProvider 能提供的速度信息的要求
setCostAllowed(boolean costAllowed)	设置 LocationProvider 是否免费
setPowerRequirement(int level)	设置 LocationProvider 的耗电量

在例 12.1 中，可以通过 Criteria 类设置 GPS 能够提供的海拔、方向、速度信息，再通过 getBestProvider()方法获得最佳的 LocationProvider，相关代码如下：

```
Criteria criteria = new Criteria();
criteria.setAccuracy(Criteria.ACCURACY_FINE);
criteria.setAltitudeRequired(true);              // 要求能提供海拔信息
criteria.setBearingRequired(true);               // 要求能提供方向信息
criteria.setCostAllowed(false);                  // 要求是免费的
criteria.setPowerRequirement(Criteria.POWER_LOW); // 要求是低功耗的
// 获取最佳的 LocationProvider
String provider = lm.getBestProvider(criteria, true);
```

12.3 获取定位信息

在了解了 GPS 的相关接口之后，可以编写一个小程序，用于获取定位信息。在本例中，实现了初始化 GPS、获取经/纬度、海拔、速度等信息。

【例 12.2】 通过 GPS 获取定位信息实例

代码如下：

```java
package com.example.location;
import android.content.Context;
import android.content.Intent;
import android.location.Location;
import android.location.LocationListener;
import android.location.LocationManager;
import android.provider.Settings;
import android.support.v7.app.AppCompatActivity;
import android.os.Bundle;
import android.widget.TextView;
import android.widget.Toast;

public class MainActivity extends AppCompatActivity {
    private TextView tvLag,tvAlt,tvSpeed;
    private LocationManager lm = null;
    private Location mLocation;
    private MyLocationListner mLocationListner;
    @Override
    protected void onCreate(Bundle savedInstanceState) {
        super.onCreate(savedInstanceState);
        setContentView(R.layout.activity_main);
        tvLag = (TextView) findViewById(R.id.tvLag);
        tvAlt = (TextView) findViewById(R.id.tvAlt);
        tvSpeed = (TextView) findViewById(R.id.tvSpeed);
        lm = (LocationManager)getSystemService(Context.LOCATION_SERVICE);
        initLocation();
    }
    private void initLocation(){
        // 判断 GPS 是否正常启动
        if(!lm.isProviderEnabled(LocationManager.GPS_PROVIDER)){
            Toast.makeText(MainActivity.this, "请开启GPS…",
                    Toast.LENGTH_SHORT);
            // 返回开启 GPS 导航的设置界面
            Intent intent = new Intent(Settings
                    .ACTION_LOCATION_SOURCE_SETTINGS);
            startActivityForResult(intent,0);
            return;
        }
        if (mLocationListner == null) {
            mLocationListner = new MyLocationListner();
        }
        // 需要注意的是，与 Eclipse 不同，AndroidSdudio 加入了运行时权限机制，需要用
```

```java
    // try/catch 抛出异常
    try{
        mLocation = lm.getLastKnownLocation(lm.GPS_PROVIDER);
        updateView(mLocation);
    }catch (SecurityException se){ }
    try{
        lm.requestLocationUpdates(LocationManager.GPS_PROVIDER,
            3000, 1, mLocationListner);
    }catch (SecurityException se){
    }
}
private class MyLocationListner implements LocationListener
{
    @Override
    public void onLocationChanged(Location location){
        updateView(location); }
    @Override
    public void onProviderDisabled(String provider){
        updateView(null); }
    @Override
    public void onProviderEnabled(String provider) {
        try{
            updateView(lm.getLastKnownLocation(provider));
        }catch (SecurityException e){}}
    @Override
    public void onStatusChanged(String provider, int status, Bundle extras){
        }
}
private void updateView(Location location)
{
    if (location!=null) {
        tvLag.setText("当前经/纬度：" + location.getLongitude() + "," +
            location.getLatitude());
        tvAlt.setText("当前海拔：" + location.getAltitude() + "m");
        tvSpeed.setText("当前速度：" + location.getSpeed() + "m/s");
    }else{
        tvLag.setText("当前经/纬度：");
        tvAlt.setText("当前海拔："  );
        tvSpeed.setText("当前速度：");
    }}}
```

在 Genymotion 模拟器中，免费提供了 GPS 设置功能，可以很方便地设置经/纬度信息与海拔信息。在 Genymotion 模拟器中开启 GPS，如图 12.2 所示，手动输入经/纬度信息，程序运行效果如图 12.3 所示。

图 12.2 在 Genymotion 模拟器中开启 GPS

图 12.3 程序运行效果

第 2 篇
Android 实验篇

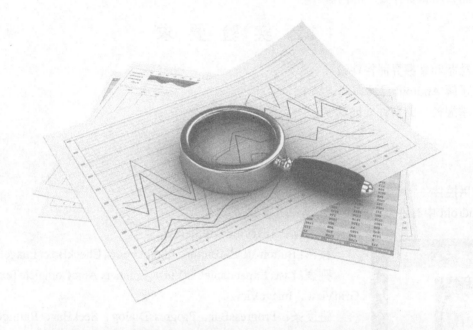

实验 1　简单 UI 设计
实验 2　高级 UI 设计
实验 3　Intent 与 Activity 的使用
实验 4　Android 资源访问
实验 5　图形、图片与多媒体
实验 6　Android 网络编程基础
实验 7　SQLite 和 SQLiteDatabase 的使用

实验 1　简单 UI 设计

1.1　实验目的

本次实验的目的是让大家熟悉 Android 开发中的 UI 设计，包括了解和熟悉常用控件的使用、界面布局和事件处理等内容。

1.2　实验要求

- 熟悉和掌握界面控件设计。
- 了解 Android 界面布局。
- 掌握控件的事件处理功能。

1.3　实验内容

常用控件

Android 中有许多常用控件（简单分类）。

图 E1.1　运行效果

文本框：TextView、EditText。
按钮：Button、RadioButton、RadioGroup、CheckBox、ImageButton。
列表：List、ExpandableListView、Spinner、AutoCompleteTextView、GridView、ImageView。
进度条：ProgressBar、ProgressDialog、SeekBar、RatingBar。
选择器：DatePicker、TimePicker。
菜单：Menu、ContentMenu。
对话框：Dialog、ProgressDialog。
常用的控件有文本框、按钮和列表等。

【练习 1.1】运用 ImageView

实现运用 ImageView 显示带边框的图片。

- 运行效果如图 E1.1 所示。
- 资源文件布局如图 E1.2 所示。

图 E1.2　资源文件布局

实验 1 简单 UI 设计

- activity_main.xml 文件代码如下：

```xml
<?xml version="1.0" encoding="utf-8"?>
<RelativeLayout xmlns:android="http://schemas.android.com/apk/res/android"
    xmlns:tools="http://schemas.android.com/tools"
    android:layout_width= "match_parent"
    android:layout_height="match_parent"
    android:paddingLeft="@dimen/activity_ horizontal_margin"
    android:paddingRight="@dimen/activity_horizontal_margin"
    android:paddingTop="@dimen/activity_vertical_margin"
    android:paddingBottom="@dimen/activity_vertical_margin"
    tools:context= ".MainActivity">

    <ImageView
        android:id="@+id/imageView1"
        android:padding="2dp"
        android:layout_margin="10dp"
        android:layout_width="wrap_content"
        android:layout_height="wrap_content"
        android:background="#000"
        android:src="@drawable/ic_launcher" />

</RelativeLayout>
```

其中，"android:src="@drawable/ic_launcher""为设置图片资源的语句。

【练习 1.2】运用 CheckBox

实现在勾选复选框后，"开始"按钮才可用，否则"开始"按钮为不可用状态。

- 运行效果如图 E1.3 所示。
- 资源文件布局如图 E1.4 所示。

图 E1.3　运行效果　　　　　　　　图 E1.4　资源文件布局

- activity_main.xml 文件代码如下：

```xml
<?xml version="1.0" encoding="utf-8"?>
<RelativeLayout xmlns:android="http://schemas.android.com/apk/res/android"
    xmlns:tools="http://schemas.android.com/tools"
    android:layout_width= "match_parent"
    android:layout_height="match_parent"
    android:paddingLeft="@dimen/activity_ horizontal_margin"
    android:paddingRight="@dimen/activity_horizontal_margin"
    android:paddingTop="@dimen/activity_vertical_margin"
    android:paddingBottom="@dimen/activity_vertical_margin"
    tools:context= ".MainActivity">

<CheckBox
    android:id="@+id/checkBox1"
    android:layout_width="wrap_content"
    android:layout_height="wrap_content"
    android:text="让按钮可用" />
<Button
    android:id="@+id/button1"
    android:layout_width="wrap_content"
    android:layout_height="wrap_content"
    android:enabled="false"
    android:text="开始"
    android:layout_below="@+id/checkBox1"
    android:layout_alignParentStart="true"
    android:layout_alignEnd="@+id/checkBox1" />

</RelativeLayout>
```

- MainActivity.java 文件代码如下:

```java
package com.example.checkbox;

import android.app.Activity;
import android.os.Bundle;
import android.view.View;
import android.widget.Button;
import android.widget.CheckBox;
import android.widget.CompoundButton;
import android.widget.Toast;

public class MainActivity extends Activity {

    @Override
    protected void onCreate(Bundle savedInstanceState) {
        super.onCreate(savedInstanceState);
        setContentView(R.layout.activity_main);
        final Button button=(Button)findViewById(R.id.button1);
        button.setOnClickListener(new View.OnClickListener() {
            @Override
            public void onClick(View v) {
```

```
                Toast.makeText(MainActivity.this, "单击了按钮！", Toast.
LENGTH_SHORT).show();

                    }
                });
                CheckBox checkbox=(CheckBox)findViewById(R.id.checkBox1);
                checkbox.setOnCheckedChangeListener(new CompoundButton.
OnCheckedChangeListener() {
                    @Override
                    public void onCheckedChanged(CompoundButton buttonView, boolean
isChecked) {
                        if (isChecked) {            // 判断该复选框是否被勾选
                            button.setEnabled(true);    // 设置按钮可用
                        } else {
                            button.setEnabled(false);   // 设置按钮不可用
                        }
                    }
                });

    }
}
```

【练习 1.3】运用 ListView

编写 Android 程序，实现图标在上、文字在下的 ListView 展示效果。
- 运行效果如图 E1.5 所示。
- 资源文件布局如图 E1.6 所示。

图 E1.5 运行效果

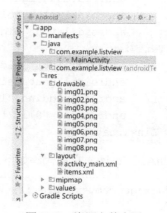

图 E1.6 资源文件布局

- 图标文件（img01.png～img08.png）如图 E1.7 所示。

图 E1.7 图标文件

- activity_main.xml 文件代码如下：

```xml
<?xml version="1.0" encoding="utf-8"?>
<LinearLayout xmlns:android="http://schemas.android.com/apk/res/android"
    android:orientation="vertical"
    android:layout_width="fill_parent"
    android:layout_height="fill_parent">
<ListView
        android:id="@+id/listView1"
        android:layout_height="wrap_content"
        android:layout_width="match_parent"/>

</LinearLayout>
```

- items.xml 文件代码如下：

```xml
<?xml version="1.0" encoding="utf-8"?>
<LinearLayout
    xmlns:android="http://schemas.android.com/apk/res/android"
    android:orientation="vertical"

    android:layout_width="match_parent"
    android:layout_height="match_parent">
<ImageView
        android:id="@+id/image"
        android:adjustViewBounds="true"
        android:maxWidth="72dp"
        android:maxHeight="72dp"
        android:layout_height="wrap_content"
        android:layout_width="wrap_content"/>
<TextView
        android:layout_width="wrap_content"
        android:layout_height="wrap_content"
        android:padding="10dp"
        android:id="@+id/title"/>
</LinearLayout>
```

- MainActivity.java 文件代码如下：

```java
package com.example.listview;

import android.app.Activity;
import android.os.Bundle;
import android.widget.ListView;
import android.widget.SimpleAdapter;
```

```java
import java.util.ArrayList;
import java.util.HashMap;
import java.util.List;
import java.util.Map;

public class MainActivity extends Activity {
    /** Called when the activity is first created. */
    @Override
    public void onCreate(Bundle savedInstanceState) {
        super.onCreate(savedInstanceState);
        setContentView(R.layout.activity_main);
        // 获取列表视图
        ListView listview = (ListView) findViewById(R.id.listView1);
        // 定义并初始化保存图片 ID 的数组
        int[] imageId = new int[] { R.drawable.img01, R.drawable.img02,
                R.drawable.img03, R.drawable.img04, R.drawable.img05,
                R.drawable.img06, R.drawable.img07, R.drawable.img08 };
        // 定义并初始化保存列表项文字的数组
        String[] title = new String[] { "保密设置", "安全", "系统设置", "上网",
"我的文档","GPS 导航", "我的音乐", "E-mail" };
        // 创建一个 List 集合
        List<Map<String, Object>> listItems = new ArrayList<Map<String, Object>>();
        // 通过 for 循环将图片 ID 和列表项文字放到 Map 中,并添加到 List 集合中
        for (int i = 0; i < imageId.length; i++) {
            // 实例化 Map 对象
            Map<String, Object> map = new HashMap<String, Object>();
            map.put("image", imageId[i]);
            map.put("title", title[i]);
            listItems.add(map);                 // 将 Map 对象添加到 List 集合中
        }

        SimpleAdapter adapter = new SimpleAdapter(this, listItems,
                R.layout.items, new String[] { "title", "image" }, new int[] {
                R.id.title, R.id.image });      // 创建 SimpleAdapter
        listview.setAdapter(adapter);           // 将适配器与 ListView 关联


    }
}
```

1.4 实 验 报 告

1. 每人一份实验报告,统一使用学校提供的 A4 尺寸的实验报告册书写或者使用 A4 尺寸的纸打印。如果需要打印,则必须采用如表 E1.1 所示的表头格式。

表 E1.1 表头格式

实验课程								
实验名称								
实验时间	学年		学期	周 星期		第 节		
学生姓名			学号			班级		
同组姓名			学号			班级		
实验地点			设备号			指导教师		

2．实验内容的主要结果及对结果的分析。
3．实验过程中所遇到的问题的解决办法。
4．心得体会、意见和建议。

1.5　实验成绩考核

1．考勤方面的得分占总分的 10%。
2．相互协作完成实验任务方面的得分占总分的 40%。
3．实验报告方面的得分占总分的 50%。

实验 2 高级 UI 设计

2.1 实验目的

本次实验的目的是让大家熟悉 Android 开发中的高级 UI 设计，包括了解和熟悉常用高级控件，以及消息提示框、对话框等内容。通过使用这些控件，用户可以开发出更优秀的用户界面。

2.2 实验要求

- 熟悉和掌握界面高级控件的设计和使用。
- 掌握进度条的用法。
- 掌握 GridView 的用法。
- 掌握 AltertDialog 的用法。

2.3 实验内容

常用控件

Android 高级控件如下所述。

进度条：ProgressBar、ProgressDialog、SeekBar、RatingBar。

选择器：DatePicker、TimePicker。

菜单：Menu、ContentMenu。

对话框：Dialog、ProgressBar、AltertDialog。

【练习 2.1】运用进度条

在主页面上设计一个按钮，并在单击按钮之后开始线程的周期，使其在运行的过程中显示 ProgressBar。

- 运行效果如图 E2.1 所示。

图 E2.1 运行效果

- 资源文件布局如图 E2.2 所示。

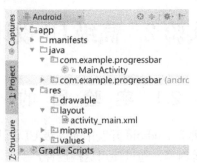

图 E2.2　资源文件布局

- activity_main.xml 文件代码如下：

```xml
<?xml version="1.0" encoding="utf-8"?>
<RelativeLayout xmlns:android="http://schemas.android.com/apk/res/android"
    xmlns:tools="http://schemas.android.com/tools"
    android:layout_width= "match_parent"
    android:layout_height="match_parent"
    android:paddingLeft="@dimen/activity_ horizontal_margin"
    android:paddingRight="@dimen/activity_horizontal_margin"
    android:paddingTop="@dimen/activity_vertical_margin"
    android:paddingBottom="@dimen/activity_vertical_margin"
    tools:context= ".MainActivity">

<TextView
        android:layout_width="wrap_content"
        android:layout_height="wrap_content"
        android:id="@+id/textView"
        android:layout_alignParentTop="true"
        android:layout_centerHorizontal="true"
        android:textSize="30dp"
        android:text="Progress bar" />

<Button
        android:layout_width="wrap_content"
        android:layout_height="wrap_content"
        android:text="Download"
        android:onClick="download"
        android:id="@+id/button2"
        android:layout_marginLeft="125dp"
        android:layout_marginStart="125dp"
        android:layout_centerVertical="true" />
</RelativeLayout>
```

- MainActivity.java 文件代码如下：

```java
package com.example.progressbar;
```

```java
import android.app.ProgressDialog;
import android.os.Bundle;
import android.support.v7.app.AppCompatActivity;
import android.view.View;
import android.widget.Button;

public class MainActivity extends AppCompatActivity {
    Button b1;
    private ProgressDialog progress;
    @Override
    protected void onCreate(Bundle savedInstanceState) {
        super.onCreate(savedInstanceState);
        setContentView(R.layout.activity_main);
        b1 = (Button) findViewById(R.id.button2);
    }
    public void download(View view){
        progress=new ProgressDialog(this);
        progress.setMessage("Downloading Music");
        progress.setProgressStyle(ProgressDialog.STYLE_HORIZONTAL);
        progress.setIndeterminate(true);
        progress.setProgress(0);
        progress.show();

        final int totalProgressTime = 100;
        final Thread t = new Thread() {
            @Override
            public void run() {
                int jumpTime = 0;

                while(jumpTime < totalProgressTime) {
                    try {
                        sleep(200);
                        jumpTime += 5;
                        progress.setProgress(jumpTime);
                    }
                    catch (InterruptedException e) {
                        // TODO Auto-generated catch block
                        e.printStackTrace();
                    }
                }
            }
        };
        t.start();
    }
}
```

【练习 2.2】运用 GridView

编写 Android 程序，实现带预览的图片浏览器。

- 运行效果如图 E2.3 所示。

- 资源文件布局如图 E2.4 所示。

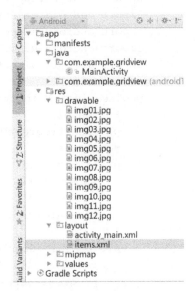

图 E2.3　运行效果　　　　　　　　图 E2.4　资源文件布局

- activity_main.xml 文件代码如下：

```xml
<?xml version="1.0" encoding="utf-8"?>
<LinearLayout xmlns:android="http://schemas.android.com/apk/res/android"
    android:orientation="horizontal"
    android:layout_width="fill_parent"
    android:layout_height="fill_parent"
    android:id="@+id/llayout">

<GridView android:id="@+id/gridView1"
        android:layout_height="match_parent"
        android:layout_width="640px"
        android:layout_marginTop="10px"
        android:horizontalSpacing="3px"
        android:verticalSpacing="3px"
        android:numColumns="4"/>

<!-- 添加一个图片切换器 -->
<ImageSwitcher
        android:id="@+id/imageSwitcher1"
        android:padding="20px"
        android:layout_width="match_parent"
        android:layout_height="match_parent"/>
</LinearLayout>
```

- items.xml 文件代码如下：

```xml
<?xml version="1.0" encoding="utf-8"?>
<LinearLayout
    xmlns:android="http://schemas.android.com/apk/res/android"
```

```xml
        android:orientation="vertical"
        android:gravity="center_horizontal"
        android:layout_width="match_parent"
        android:layout_height="match_parent">
<ImageView
        android:id="@+id/image"
        android:paddingLeft="10px"
        android:paddingTop="20px"
        android:paddingBottom="20px"
        android:adjustViewBounds="true"
        android:maxWidth="300px"
        android:maxHeight="226px"
        android:layout_height="wrap_content"
        android:layout_width="wrap_content"/>
<TextView
        android:layout_width="wrap_content"
        android:layout_height="wrap_content"
        android:layout_gravity="center"
        android:id="@+id/title" />
</LinearLayout>
```

- **MainActivity.java 文件代码如下：**

```java
package com.example.gridview;

import android.app.Activity;
import android.os.Bundle;
import android.view.View;
import android.view.ViewGroup.LayoutParams;
import android.view.animation.AnimationUtils;
import android.widget.AdapterView;
import android.widget.AdapterView.OnItemClickListener;
import android.widget.GridView;
import android.widget.ImageSwitcher;
import android.widget.ImageView;
import android.widget.SimpleAdapter;
import android.widget.ViewSwitcher.ViewFactory;

import java.util.ArrayList;
import java.util.HashMap;
import java.util.List;
import java.util.Map;

public class MainActivity extends Activity {
    // 定义并初始化保存图片 ID 的数组
    private int[] imageId = new int[] { R.drawable.img01, R.drawable.img02,
            R.drawable.img03, R.drawable.img04, R.drawable.img05,
            R.drawable.img06, R.drawable.img07, R.drawable.img08,
            R.drawable.img09, R.drawable.img10, R.drawable.img11,
            R.drawable.img12, };
```

```java
        // 定义并初始化保存列表项文字的数组
        final String[] filename = new String[] { "img01.jpg", "img02.jpg",
"img03.jpg","img04.jpg","img05.jpg","img06.jpg","img07.jpg","img08.jpg","img09
.jpg", "img10.jpg","img11.jpg","img12.jpg" };

        private ImageSwitcher imageSwitcher;    // 声明一个图片切换器对象
        @Override
        public void onCreate(Bundle savedInstanceState) {
            super.onCreate(savedInstanceState);
            setContentView(R.layout.activity_main);
            /*使用SimpleAdapter指定要显示的内容*/
            // 创建一个List集合
            List<Map<String, Object>> listItems = new ArrayList<Map<String,
Object>>();
            // 通过for循环将图片ID和列表项文字放到Map中,并添加到List集合中
            for (int i = 0; i < imageId.length; i++) {
                // 实例化Map对象
                Map<String, Object> map = new HashMap<String, Object>();
                map.put("image", imageId[i]);
                map.put("title", filename[i]);
                listItems.add(map);               // 将Map对象添加到List集合中
            }

            final SimpleAdapter adapter = new SimpleAdapter(this, listItems,
                    R.layout.items, new String[] { "title", "image" }, new int[] {
                    R.id.title, R.id.image });    // 创建SimpleAdapter
            // 获取图片切换器
            imageSwitcher = (ImageSwitcher) findViewById(R.id.imageSwitcher1);
            // 设置动画效果
            imageSwitcher.setInAnimation(AnimationUtils.loadAnimation(this,
                    android.R.anim.fade_in));     // 设置淡入动画
            imageSwitcher.setOutAnimation(AnimationUtils.loadAnimation(this,
                    android.R.anim.fade_out));    // 设置淡出动画
            imageSwitcher.setFactory(new ViewFactory() {

                @Override
                public View makeView() {
                    // 实例化一个ImageView类的对象
                    ImageView imageView = new ImageView(MainActivity.this);
                    // 设置保持纵横比居中缩放图片
                    imageView.setScaleType(ImageView.ScaleType.FIT_CENTER);
                    imageView.setLayoutParams(new ImageSwitcher.LayoutParams(
                            LayoutParams.WRAP_CONTENT, LayoutParams.WRAP_CONTENT));
                    return imageView;             // 返回imageView对象
                }

            });
            imageSwitcher.setImageResource(imageId[6]); // 设置默认显示的图片
            // 获取GridView
```

```
            GridView gridview = (GridView) findViewById(R.id.gridView1);
            gridview.setAdapter(adapter);        // 将适配器与 GridView 关联
            gridview.setOnItemClickListener(new OnItemClickListener() {

                @Override
                public void onItemClick(AdapterView<?> parent, View view, int position,
                                long id) {
                    // 显示选中的图片
                    imageSwitcher.setImageResource(imageId[position]);

                }

            });
        }
    }
```

【练习 2.3】运用 AltertDialog

编写 Android 程序，运用 AltertDialog 实现自定义的登录对话框。
- 运行效果如图 E2.5 所示。
- 资源文件布局如图 E2.6 所示。

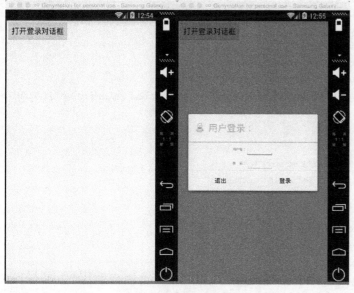

图 E2.5 运行效果   图 E2.6 资源文件布局

- activity_main.xml 文件代码如下：

```xml
<?xml version="1.0" encoding="utf-8"?>
<LinearLayout xmlns:android="http://schemas.android.com/apk/res/android"
    android:orientation="vertical"
    android:layout_width="fill_parent"
    android:layout_height="fill_parent">
<Button
```

```
            android:id="@+id/button1"
            android:layout_width="wrap_content"
            android:layout_height="wrap_content"
            android:text="打开登录对话框" />
    </LinearLayout>
```

- login.xml 文件代码如下：

```
<?xml version="1.0" encoding="utf-8"?>
<TableLayout android:id="@+id/tableLayout1"
    android:layout_width="fill_parent"
    android:layout_height="fill_parent"
    xmlns:android="http://schemas.android.com/apk/res/android"
    android:gravity="center_vertical"
    android:stretchColumns="0,3">
<!-- 第一行 -->
<TableRow android:id="@+id/tableRow1"
        android:layout_width="wrap_content"
        android:layout_height="wrap_content">
<TextView/>
<TextView android:text="用户名："
        android:id="@+id/textView1"
        android:layout_width="wrap_content"
        android:textSize="24px"
        android:layout_height="wrap_content"/>
<EditText android:id="@+id/editText1"
        android:textSize="24px"
        android:layout_width="wrap_content"
        android:layout_height="wrap_content" android:minWidth="200px"/>
<TextView/>
</TableRow>
<!-- 第二行 -->
<TableRow android:id="@+id/tableRow2"
        android:layout_width="wrap_content"
        android:layout_height="wrap_content">
<TextView/>
<TextView android:text="密码："
        android:id="@+id/textView2"
        android:textSize="24px"
        android:layout_width="wrap_content"
        android:layout_height="wrap_content"/>
<EditText android:layout_height="wrap_content"
        android:layout_width="wrap_content"
        android:textSize="24px"
        android:id="@+id/editText2"
        android:inputType="textPassword"/>
<TextView/>
</TableRow>
</TableLayout>
```

- MainActivity.java 文件代码如下：

```java
package com.example.altertdialog;

import android.app.Activity;
import android.app.AlertDialog;
import android.app.AlertDialog.Builder;
import android.os.Bundle;
import android.view.LayoutInflater;
import android.view.View;
import android.widget.Button;

public class MainActivity extends Activity {
    @Override
    public void onCreate(Bundle savedInstanceState) {
        super.onCreate(savedInstanceState);
        setContentView(R.layout.activity_main);


        // 自定义的用户登录对话框
        // 获取布局文件中添加的按钮
        Button button1 = (Button) findViewById(R.id.button1);
        button1.setOnClickListener(new View.OnClickListener() {

            @Override
            public void onClick(View v) {
                Builder builder = new AlertDialog.Builder(MainActivity.this);
                builder.setIcon(R.drawable.advise);          // 设置对话框的图标
                builder.setTitle("用户登录：");               // 设置对话框的标题
                LayoutInflater inflater = getLayoutInflater();
                View view = inflater.inflate(R.layout.login, null);
                builder.setView(view);
                builder.setPositiveButton("登录", null); // 添加"登录"按钮
                builder.setNegativeButton("退出", null); // 添加"退出"按钮
                builder.create().show();                      // 创建对话框并显示
            }
        });

    }
}
```

2.4 实验报告

1. 每人一份实验报告，统一使用学校提供的 A4 尺寸的实验报告册书写或者使用 A4 尺寸的纸打印。如果需要打印，则必须采用如表 E2.1 所示的表头格式。

表 E2.1　表头格式

实验课程							
实验名称							
实验时间	学年		学期	周 星期		第　节	
学生姓名		学号			班级		
同组姓名		学号			班级		
实验地点			设备号			指导教师	

2. 实验内容的主要结果及对结果的分析。
3. 实验过程中所遇到的问题的解决办法。
4. 心得体会、意见和建议。

2.5　实验成绩考核

1. 考勤方面的得分占总分的 10%。
2. 相互协作完成实验任务方面的得分占总分的 40%。
3. 实验报告方面的得分占总分的 50%。

实验 3　Intent 与 Activity 的使用

3.1　实　验　目　的

本次实验的目的是让大家熟悉 Intent 和 Activity 的使用。Intent 通常用于绑定应用程序组件。Intent 的作用是在应用程序中 Activity 之间进行启动、停止和传输等操作，并实现添加用户名、密码小程序。

3.2　实　验　要　求

- 掌握创建、配置、启动和关闭 Activity 的方法。
- 掌握如何使用 Bundle 在 Activity 之间交换数据。
- 掌握 Intent 对象的使用。

3.3　实　验　内　容

【练习 3.1】从一个 Activity 跳到另一个 Activity

编写 Android 程序，实现根据输入的阳历生日判断相应用户的星座。
- 运行效果如图 E3.1 所示。
- 资源文件布局如图 E3.2 所示。

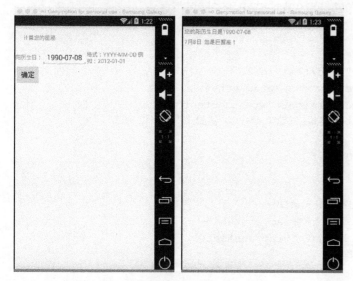

图 E3.1　运行效果

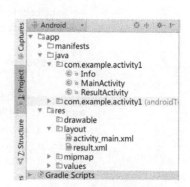

图 E3.2　资源文件布局

- activity_main.xml 文件代码如下：

```xml
<?xml version="1.0" encoding="utf-8"?>
<LinearLayout xmlns:android="http://schemas.android.com/apk/res/android"
    android:layout_width="fill_parent"
    android:layout_height="fill_parent"
    android:orientation="vertical" >

    <TextView
        android:layout_width="fill_parent"
        android:layout_height="wrap_content"
        android:layout_gravity="center_horizontal"
        android:padding="20dp"
        android:text="计算您的星座" />

    <LinearLayout
        android:id="@+id/linearLayout1"
        android:gravity="center_vertical"
        android:layout_width="match_parent"
        android:layout_height="wrap_content" >

        <TextView
            android:id="@+id/textView1"
            android:layout_width="wrap_content"
            android:layout_height="wrap_content"
            android:text="阳历生日：" />

        <EditText
            android:id="@+id/birthday"
            android:minWidth="100dp"
            android:layout_width="wrap_content"
            android:layout_height="wrap_content">
        </EditText>

        <TextView
            android:id="@+id/textView2"
            android:layout_width="wrap_content"
            android:layout_height="wrap_content"
            android:text="格式：YYYY-MM-DD 例如：2012-01-01" />

    </LinearLayout>

    <Button
        android:id="@+id/button1"
        android:layout_width="wrap_content"
        android:layout_height="wrap_content"
        android:text="确定" />

</LinearLayout>
```

- result.xml 文件代码如下：

```xml
<?xml version="1.0" encoding="utf-8"?>
<LinearLayout xmlns:android="http://schemas.android.com/apk/res/android"
    android:layout_width="match_parent"
    android:layout_height="match_parent"
    android:orientation="vertical" >

    <TextView
        android:id="@+id/birthday"
        android:layout_width="wrap_content"
        android:layout_height="wrap_content"
        android:padding="10px"
        android:text="阳历生日" />

    <TextView
        android:id="@+id/result"
        android:padding="10px"
        android:layout_width="wrap_content"
        android:layout_height="wrap_content"
        android:text="星座" />
</LinearLayout>
```

- MainActivity.java 文件代码如下：

```java
package com.example.activity1;

import android.app.Activity;
import android.content.Intent;
import android.os.Bundle;
import android.view.View;
import android.view.View.OnClickListener;
import android.widget.Button;
import android.widget.EditText;
import android.widget.Toast;

public class MainActivity extends Activity {
    @Override
    public void onCreate(Bundle savedInstanceState) {
        super.onCreate(savedInstanceState);
        setContentView(R.layout.activity_main);
        Button button=(Button)findViewById(R.id.button1);
        button.setOnClickListener(new OnClickListener() {

            @Override
            public void onClick(View v) {
                Info info=new Info();   // 实例化一个保存输入的基本信息的对象

                if("".equals(((EditText)findViewById(R.id.birthday)).getText().toString())){
```

```
                        Toast.makeText(MainActivity.this,"请输入您的阳历生日,否则
不能计算!", Toast.LENGTH_SHORT).show();
                        return;
                    }
                    String  birthday=((EditText)findViewById(R.id.birthday)).
getText().toString();

                    info.setBirthday(birthday);      // 设置生日
                    Bundle bundle=new Bundle();      // 实例化一个 Bundle 对象
                    // 将输入的基本信息保存到 Bundle 对象中
                    bundle.putSerializable("info", info);
                    Intent intent=new Intent(MainActivity.this,ResultActivity.class);
                    intent.putExtras(bundle);        // 将 bundle 保存到 Intent 对象中
                    startActivity(intent);           // 启动 intent 对应的 Activity
                }
            });
        }
    }
```

- Info.java 文件代码如下:

```
package com.example.activity1;

import java.io.Serializable;

public class Info implements Serializable {

    private static final long serialVersionUID = 1L;
    private String birthday="";       // 生日
    public String getBirthday() {
        return birthday;
    }
    public void setBirthday(String birthday) {
        this.birthday = birthday;
    }

}
```

- ResultActivity.java 文件代码如下:

```
package com.example.activity1;

import android.app.Activity;
import android.content.Intent;
import android.os.Bundle;
import android.widget.TextView;

public class ResultActivity extends Activity {
```

```java
@Override
protected void onCreate(Bundle savedInstanceState) {
    super.onCreate(savedInstanceState);
    setContentView(R.layout.result);              // 设置该Activity使用的布局
    // 获取显示生日的文本框
    TextView birthday = (TextView) findViewById(R.id.birthday);
    // 获取显示星座的文本框
    TextView result = (TextView) findViewById(R.id.result);
    Intent intent = getIntent();                  // 获取Intent对象
    Bundle bundle = intent.getExtras();           // 获取传递的数据包
    // 获取一个可序列化的Info对象
    Info info = (Info) bundle.getSerializable("info");
    // 获取性别并显示到相应文本框中
    birthday.setText("您的阳历生日是" + info.getBirthday());

    result.setText( query(info.getBirthday()));   // 显示计算后的星座
}

/**
 * 功能：根据生日查询星座
 *
 * param month
 * param day
 * return
 */
public String query(String birthday) {
    int month=0;
    int day=0;
    try{
        month=Integer.parseInt(birthday.substring(5, 7));
        day=Integer.parseInt(birthday.substring(8, 10));
    }catch(Exception e){
        e.printStackTrace();
    }
    String name = "";                             // 提示信息
    // 如果输入的月和日有效
    if (month > 0 && month < 13 && day > 0 && day < 32) {
        if ((month == 3 && day > 20) || (month == 4 && day < 21)) {
            name = "您是白羊座！";
        } else if ((month == 4 && day > 20) || (month == 5 && day < 21)) {
            name = "您是金牛座！";
        } else if ((month == 5 && day > 20) || (month == 6 && day < 22)) {
            name = "您是双子座！";
        } else if ((month == 6 && day > 21) || (month == 7 && day < 23)) {
            name = "您是巨蟹座！";
        } else if ((month == 7 && day > 22) || (month == 8 && day < 23)) {
            name = "您是狮子座！";
        } else if ((month == 8 && day > 22) || (month == 9 && day < 23)) {
            name = "您是处女座！";
```

```
        } else if ((month == 9 && day > 22) || (month == 10 && day < 23)) {
            name = "您是天秤座！";
        } else if ((month == 10 && day > 22) || (month == 11 && day < 22)) {
            name = "您是天蝎座！";
        } else if ((month == 11 && day > 21) || (month == 12 && day < 22)) {
            name = "您是射手座！";
        } else if ((month == 12 && day > 21) || (month == 1 && day < 20)) {
            name = "您是摩羯座！";
        } else if ((month == 1 && day > 19) || (month == 2 && day < 19)) {
            name = "您是水牛座！";
        } else if ((month == 2 && day > 18) || (month == 3 && day < 21)) {
            name = "您是双鱼座！";
        }
        name = month + "月" + " " + day + "日  " + name;
    } else {                                       // 如果输入的月和日无效
        name = "您输入的生日格式不正确或者不是真实生日！";
    }
    return name;                                   // 返回星座或提示信息
}
```

- AndroidManifest.xml 文件代码如下：

```xml
<?xml version="1.0" encoding="utf-8"?>
<manifest xmlns:android="http://schemas.android.com/apk/res/android"
    package="com.example.activity1" >

    <application
        android:allowBackup="true"
        android:icon="@mipmap/ic_launcher"
        android:label="@string/app_name"
        android:supportsRtl="true"
        android:theme="@style/AppTheme" >
        <activity android:name=".MainActivity" >
            <intent-filter>
                <action android:name="android.intent.action.MAIN" />

                <category android:name="android.intent.category.LAUNCHER" />
            </intent-filter>
        </activity>
        <activity android:name=".ResultActivity" >
        </activity>
    </application>

</manifest>
```

【练习 3.2】从一个 Activity 跳到另一个 Activity 再返回

编写 Android 程序，实现需选择所在城市的用户注册。

- 运行效果如图 E3.3 所示。

实验 3　Intent 与 Activity 的使用

- 资源文件布局如图 E3.4 所示。

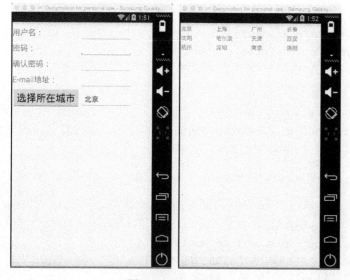

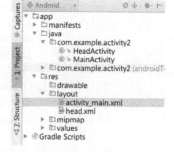

图 E3.3　运行效果　　　　　　　　　　图 E3.4　资源文件布局

- activity_main.xml 文件代码如下：

```
<?xml version="1.0" encoding="utf-8"?>
<LinearLayout xmlns:android="http://schemas.android.com/apk/res/android"
    android:layout_width="fill_parent"
    android:layout_height="fill_parent"
    android:orientation="horizontal"
    android:paddingTop="20px" >

    <TableLayout
        android:id="@+id/tableLayout1"
        android:layout_width="match_parent"
        android:layout_height="wrap_content" >

        <TableRow
            android:id="@+id/tableRow1"
            android:layout_width="wrap_content"
            android:layout_height="wrap_content" >

            <TextView
                android:id="@+id/textView1"
                android:layout_width="wrap_content"
                android:layout_height="wrap_content"
                android:text="用户名："
                android:textSize="60px" />

            <EditText
                android:id="@+id/user"
                android:layout_width="wrap_content"
                android:layout_height="wrap_content"
                android:minWidth="400px" />
```

```xml
        </TableRow>

        <TableRow
            android:id="@+id/tableRow2"
            android:layout_width="wrap_content"
            android:layout_height="wrap_content" >

            <TextView
                android:id="@+id/textView2"
                android:layout_width="wrap_content"
                android:layout_height="wrap_content"
                android:text="密码："
                android:textSize="60px" />

            <EditText
                android:id="@+id/pwd"
                android:layout_width="wrap_content"
                android:layout_height="wrap_content"
                android:inputType="textPassword" />
        </TableRow>

        <TableRow
            android:id="@+id/tableRow3"
            android:layout_width="wrap_content"
            android:layout_height="wrap_content" >

            <TextView
                android:id="@+id/textView3"
                android:layout_width="wrap_content"
                android:layout_height="wrap_content"
                android:text="确认密码："
                android:textSize="60px" />

            <EditText
                android:id="@+id/repwd"
                android:layout_width="wrap_content"
                android:layout_height="wrap_content"
                android:inputType="textPassword" />
        </TableRow>

        <TableRow
            android:id="@+id/tableRow4"
            android:layout_width="wrap_content"
            android:layout_height="wrap_content" >

            <TextView
                android:id="@+id/textView3"
                android:layout_width="wrap_content"
                android:layout_height="wrap_content"
                android:text="E-mail 地址："
                android:textSize="60px" />
```

```xml
        <EditText
            android:id="@+id/email"
            android:layout_width="wrap_content"
            android:layout_height="wrap_content" />
    </TableRow>

    <TableRow
        android:id="@+id/tableRow5"
        android:layout_width="wrap_content"
        android:layout_height="wrap_content" >

        <Button
            android:id="@+id/button1"
            android:layout_width="wrap_content"
            android:layout_height="wrap_content"
            android:text="选择所在城市"
            android:textSize="25dp" />

        <EditText
            android:id="@+id/city"
            android:layout_width="wrap_content"
            android:layout_height="wrap_content" />


    </TableRow>
</TableLayout>

</LinearLayout>
```

- head.xml 文件代码如下：

```xml
<?xml version="1.0" encoding="utf-8"?>
<LinearLayout xmlns:android="http://schemas.android.com/apk/res/android"
    android:layout_width="match_parent"
    android:layout_height="match_parent"
    android:orientation="vertical" >
    <GridView android:id="@+id/gridView1"
        android:layout_height="match_parent"
        android:layout_width="match_parent"
        android:layout_marginTop="10px"
        android:horizontalSpacing="3px"
        android:verticalSpacing="3px"
        android:numColumns="4"/>

</LinearLayout>
```

- MainActivity.java 文件代码如下：

```java
package com.example.activity2;

import android.app.Activity;
```

```java
import android.content.Intent;
import android.os.Bundle;
import android.view.View;
import android.view.View.OnClickListener;
import android.widget.Button;
import android.widget.TextView;

public class MainActivity extends Activity {
    /** Called when the activity is first created. */
    @Override
    public void onCreate(Bundle savedInstanceState) {
        super.onCreate(savedInstanceState);
        setContentView(R.layout.activity_main);
        Button button=(Button)findViewById(R.id.button1); // 获取选择头像按钮
        button.setOnClickListener(new OnClickListener() {

            @Override
            public void onClick(View v) {
                Intent intent=new Intent(MainActivity.this,HeadActivity.class);
                startActivityForResult(intent, 0x11);// 启动intent对应的Activity
            }
        });
    }

    @Override
    protected void onActivityResult(int requestCode, int resultCode, Intent data) {
        super.onActivityResult(requestCode, resultCode, data);
        if(requestCode==0x11 && resultCode==0x11){ // 判断是否为待处理的结果
            Bundle bundle=data.getExtras();            // 获取传递的数据包
            String city=bundle.getString("city");      // 获取选择的头像ID
            // 获取布局文件中添加的ImageView
            TextView tv=(TextView)findViewById(R.id.city);
            tv.setText(city);
        }
    }
}
```

- HeadActivity.java 文件代码如下：

```java
package com.example.activity2;

import android.app.Activity;
import android.content.Intent;
import android.os.Bundle;
import android.view.View;
import android.view.ViewGroup;
import android.widget.AdapterView;
import android.widget.AdapterView.OnItemClickListener;
import android.widget.BaseAdapter;
```

```java
            import android.widget.GridView;
            import android.widget.TextView;

            public class HeadActivity extends Activity {
                public String[] city = new String[] { "北京","上海","广州","长春","沈阳",
"哈尔滨","天津","西安","杭州","深圳","南京","洛阳" };     // 定义并初始化保存头像ID的数组
                @Override
                protected void onCreate(Bundle savedInstanceState) {
                    super.onCreate(savedInstanceState);
                    setContentView(R.layout.head);         // 设置该Activity使用的布局
                    // 获取GridView
                    GridView gridview = (GridView) findViewById(R.id.gridView1);
                    BaseAdapter adapter=new BaseAdapter() {
                        @Override
                        public View getView(int position, View convertView, ViewGroup parent) {
                            TextView tv;                   // 声明ImageView的对象
                            if(convertView==null){
                                tv=new TextView(HeadActivity.this); // 实例化ImageView的对象
//                              /*************设置图片的宽度和高度******************/
//                              imageview.setAdjustViewBounds(true);
//                              imageview.setMaxWidth(158);
//                              imageview.setMaxHeight(150);
//                              /**************************************************/
                                tv.setPadding(5, 5, 5, 5); // 设置ImageView的内边距
                            }else{
                                tv=(TextView)convertView;
                            }
                            tv.setText(city[position]);    // 为ImageView设置要显示的图片
                            return tv;                     // 返回ImageView
                        }
                        /*
                         * 功能:获得当前选项的ID
                         */
                        @Override
                        public long getItemId(int position) {
                            return position;
                        }
                        /*
                         * 功能:获得当前选项
                         */
                        @Override
                        public Object getItem(int position) {
                            return position;
                        }
                        /*
                         * 获得数量
                         */
                        @Override
```

```java
            public int getCount() {
                return city.length;
            }
        };

        gridview.setAdapter(adapter);                    // 将适配器与 GridView 关联
        gridview.setOnItemClickListener(new OnItemClickListener() {
            @Override
            public void onItemClick(AdapterView<?> parent, View view, int position,long id) {
                Intent intent=getIntent();               // 获取 Intent 对象
                Bundle bundle=new Bundle();              // 实例化要传递的数据包
                bundle.putString("city",city[position] );// 显示选中的图片
                intent.putExtras(bundle);                // 将数据包保存到 Intent 中
                // 设置返回的结果码,并返回调用该 Activity 的 Activity
                setResult(0x11,intent);
                finish();                                // 关闭当前 Activity
            }
        });

    }

}
```

- AndroidManifest.xml 文件代码如下:

```xml
<?xml version="1.0" encoding="utf-8"?>
<manifest xmlns:android="http://schemas.android.com/apk/res/android"
    package="com.example.activity2" >

    <application
        android:allowBackup="true"
        android:icon="@mipmap/ic_launcher"
        android:label="@string/app_name"
        android:supportsRtl="true"
        android:theme="@style/AppTheme" >
        <activity android:name=".MainActivity" >
            <intent-filter>
                <action android:name="android.intent.action.MAIN" />

                <category android:name="android.intent.category.LAUNCHER" />
            </intent-filter>
        </activity>
        <activity android:name=".HeadActivity" >
        </activity>
    </application>

</manifest>
```

3.4 实验报告

1. 每人一份实验报告，统一使用学校提供的 A4 尺寸的实验报告册书写或者使用 A4 尺寸的纸打印。如果需要打印，则必须采用如表 E3.1 所示的表头格式。

表 E3.1　表头格式

实验课程							
实验名称							
实验时间	学年		学期	周 星期		第　节	
学生姓名		学号			班级		
同组姓名		学号			班级		
实验地点			设备号			指导教师	

2．实验内容的主要结果及对结果的分析。
3．实验过程中所遇到的问题的解决办法。
4．心得体会、意见和建议。

3.5 实验成绩考核

1．考勤方面的得分占总分的 10%。
2．相互协作完成实验任务方面的得分占总分的 40%。
3．实验报告方面的得分占总分的 50%。

实验 4 Android 资源访问

4.1 实 验 目 的

本次实验的目的是让大家熟悉 Android 中的资源。资源指的是代码中使用的外部文件,这些文件作为应用程序的一部分,会被编译到应用程序中。

4.2 实 验 要 求

1. 掌握字符串资源、颜色资源和尺寸资源文件的定义和使用。
2. 掌握通过菜单资源定义上下文菜单和选项菜单。

4.3 实 验 内 容

【练习 4.1】为 ImageView 更换图片

在实现本程序前需要准备两张图片,分别存放在/res/drawable 目录与手机文件系统的另一个目录中。在本程序中两张图片文件的存放路径分别为/res/drawable/img01.jpg 与/data/data/com.example.resource1/img02.jpg。程序会先将/res/drawable 目录里的图片显示在 ImageView 中,再利用一个 Button,当用户单击 Button 后,将 ImageView 里的图片换成另一张存在手机文件系统里的图片。

- 运行效果如图 E4.1 所示。

图 E4.1 运行效果

实验 4　Android 资源访问

- 将 img02.jpg 导入手机文件系统。

在 Android Studio 菜单栏中选择"Tools"→"Android"→"Android Device Monitor"命令，如图 E4.2 所示。

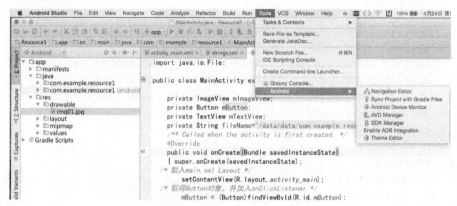

图 E4.2　选择"Android Device Monitor"命令

在弹出的"Android Device Monitor"界面中，选择"File Explorer"选项卡，找到将要导入图片的/data/data/com.example.resource1 目录，然后单击"导入"图标，导入 img02.jpg 文件，最终效果如图 E4.3 所示。

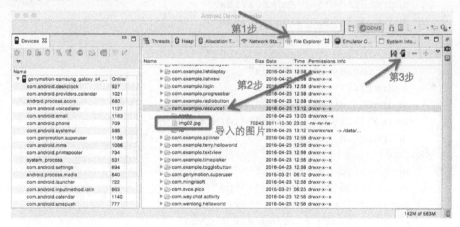

图 E4.3　最终效果

- 资源文件布局如图 E4.4 所示。

图 E4.4　资源文件布局

- activity_main.xml 文件代码如下:

```xml
<?xml version="1.0" encoding="utf-8"?>
<RelativeLayout xmlns:android="http://schemas.android.com/apk/res/android"
    xmlns:tools="http://schemas.android.com/tools"
    android:layout_width= "match_parent"
    android:layout_height="match_parent"
    android:paddingLeft="@dimen/activity_ horizontal_margin"
    android:paddingRight="@dimen/activity_horizontal_margin"
    android:paddingTop="@dimen/activity_vertical_margin"
    android:paddingBottom="@dimen/activity_vertical_margin"
    tools:context= ".MainActivity">

    <TextView
        android:id="@+id/mTextView"
        android:layout_width="fill_parent"
        android:layout_height="wrap_content"
        android:text="@string/str_text"
        android:layout_x="30px"
        android:layout_y="30px"/>
    <ImageView
        android:id="@+id/mImageView"
        android:layout_width="500px"
        android:layout_height="356px"
        android:src="@drawable/img01"
        android:layout_x="30px"
        android:layout_y="62px"
        android:layout_marginTop="87dp"
        android:layout_below="@+id/mTextView"
        android:layout_centerHorizontal="true" />
    <Button
        android:id="@+id/mButton"
        android:layout_width="155px"
        android:layout_height="wrap_content"
        android:text="@string/str_button"
        android:textSize="18sp"
        android:layout_x="80px"
        android:layout_y="302px"
        android:layout_centerVertical="true"
        android:layout_alignStart="@+id/mImageView"
        android:layout_alignEnd="@+id/mImageView" />
</RelativeLayout>
```

- strings.xml 文件代码如下:

```xml
<resources>
    <string name="app_name">Resource1</string>
    <string name="str_button">更换图片</string>
    <string name="str_text">res/drawable/img01.jpg</string>
</resources>
```

- MainActivity.java 文件代码如下：

```java
package com.example.resource1;

import android.app.Activity;
import android.graphics.Bitmap;
import android.graphics.BitmapFactory;
import android.os.Bundle;
import android.view.View;
import android.widget.Button;
import android.widget.ImageView;
import android.widget.TextView;

import java.io.File;

public class MainActivity extends Activity {

    private ImageView mImageView;
    private Button mButton;
    private TextView mTextView;
    private String fileName="/data/data/com.example.resource1/img02.jpg";
    /** Called when the activity is first created. */
    @Override
    public void onCreate(Bundle savedInstanceState)
    { super.onCreate(savedInstanceState);
        /* 载入 main.xml Layout */
        setContentView(R.layout.activity_main);
        /* 取得 Button 对象，并加入 onClickListener */
        mButton = (Button)findViewById(R.id.mButton);
        mButton.setOnClickListener(new Button.OnClickListener() {
            public void onClick(View v) {
                /* 取得对象 */
                mImageView = (ImageView) findViewById(R.id.mImageView);
                mTextView = (TextView) findViewById(R.id.mTextView);
                /* 检查文件是否存在 */
                File f = new File(fileName);
                if (f.exists()) {
                    /* 产生 Bitmap 对象，并放入 mImageView 中 */
                    Bitmap bm = BitmapFactory.decodeFile(fileName);
                    mImageView.setImageBitmap(bm);
                    mTextView.setText(fileName);
                } else {
                    mTextView.setText("文件不存在");
                }
            }
        });
    }
}
```

【练习 4.2】运用上下文菜单

编写 Android 项目，实现带子菜单的上下文菜单。

- 运行效果如图 E4.5 所示。

图 E4.5 运行效果

- 资源文件布局如图 E4.6 所示。
- 在 res 目录下创建一个 menu 目录,并在该目录中创建一个名称为 "contextmenu.xml" 的菜单资源文件。右击 res 目录,在弹出的快捷菜单中选择 "new"→"Directory"命令,创建 menu 目录;右击 menu 目录,在弹出的快捷菜单中选择"new"→"Menu resource file"命令,创建 contextmenu.xml 菜单资源文件,创建方式如图 E4.7 所示。

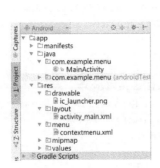

图 E4.6 资源文件布局

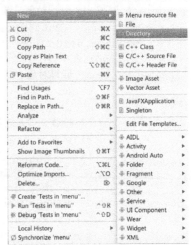

图 E4.7 创建方式

- contextmenu.xml 文件代码如下:

```xml
<?xml version="1.0" encoding="utf-8"?>
<menu xmlns:android="http://schemas.android.com/apk/res/android" >

    <item
        android:id="@+id/color1"
```

```xml
            android:title="红色">
        </item>
        <item
            android:id="@+id/color2"
            android:title="绿色">
        </item>
        <item
            android:id="@+id/color3"
            android:title="蓝色">
        </item>
        <item
            android:id="@+id/color4"
            android:title="橙色">
        </item>
        <item
            android:id="@+id/color5"
            android:title="恢复默认">
        </item>
        <item
            android:id="@+id/item2"
            android:alphabeticShortcut="e"
            android:title="其他颜色">
            <menu>
                <group
                    android:id="@+id/other" >
                    <item
                        android:id="@+id/other1"
                        android:title="橄榄色">
                    </item>
                    <item
                        android:id="@+id/other2"
                        android:title="水绿色">
                    </item>
                </group>
            </menu>
        </item>

</menu>
```

- activity_main.xml 文件代码如下：

```xml
<?xml version="1.0" encoding="utf-8"?>
<LinearLayout xmlns:android="http://schemas.android.com/apk/res/android"
    android:layout_width="fill_parent"
    android:layout_height="fill_parent"
    android:padding="5px"
    android:orientation="vertical" >
    <TextView
        android:id="@+id/show"
        android:textSize="60px"
```

```
            android:layout_width="match_parent"
            android:layout_height="wrap_content"
            android:text="打开菜单…" />

</LinearLayout>
```

- **MainActivity.java 文件代码如下：**

```java
package com.example.menu;

import android.app.Activity;
import android.graphics.Color;
import android.os.Bundle;
import android.view.ContextMenu;
import android.view.ContextMenu.ContextMenuInfo;
import android.view.MenuInflater;
import android.view.MenuItem;
import android.view.View;
import android.widget.TextView;


public class MainActivity extends Activity {
    private TextView tv;
    @Override
    public void onCreate(Bundle savedInstanceState) {
        super.onCreate(savedInstanceState);
        setContentView(R.layout.activity_main);
        tv=(TextView)findViewById(R.id.show);
        registerForContextMenu(tv);              // 为文本框注册上下文菜单

    }
    // 创建上下文菜单
    /************************************************************/
    @Override
    public void onCreateContextMenu(ContextMenu menu, View v,
                        ContextMenuInfo menuInfo) {
        // 实例化一个 MenuInflater 对象
        MenuInflater inflator=new MenuInflater(this);
        inflator.inflate(R.menu.contextmenu, menu);   // 解析菜单文件
        menu.setHeaderIcon(R.drawable.ic_launcher);   // 为菜单头设置图标
        menu.setHeaderTitle("请选择文字颜色：");        // 为菜单头设置标题

    }
    @Override
    public boolean onContextItemSelected(MenuItem item) {
        if(item.getGroupId()==R.id.other){         // 判断是否选择了参数设置菜单组
            if(item.getItemId()==R.id.other1){     // 当菜单项已经被选中
                tv.setTextColor(Color.rgb(118, 146, 60));
            }else if(item.getItemId()==R.id.other2){
                tv.setTextColor(Color.rgb(0, 255, 255));
```

```
            }
        }else{
            switch(item.getItemId()){
                case R.id.color1:      // 当选择红色时
                    tv.setTextColor(Color.rgb(255, 0, 0));
                    break;
                case R.id.color2:      // 当选择绿色时
                    tv.setTextColor(Color.rgb(0, 255, 0));
                    break;
                case R.id.color3:      // 当选择蓝色时
                    tv.setTextColor(Color.rgb(0, 0, 255));
                    break;
                case R.id.color4:      // 当选择橙色时
                    tv.setTextColor(Color.rgb(255, 180, 0));
                    break;
                default:
                    tv.setTextColor(Color.rgb(255, 255, 255));
            }
        }
        return true;
    }

}
```

4.4 实验报告

1. 每人一份实验报告，统一使用学校提供的 A4 尺寸的实验报告册书写或者使用 A4 尺寸的纸打印。如果需要打印，则必须采用如表 E4.1 所示的表头格式。

表 E4.1　表头格式

实验课程							
实验名称							
实验时间	学年	学期	周 星期	第 节			
学生姓名		学号			班级		
同组姓名		学号			班级		
实验地点		设备号			指导教师		

2. 实验内容的主要结果及对结果的分析。
3. 实验过程中所遇到的问题的解决办法。
4. 心得体会、意见和建议。

4.5 实验成绩考核

1. 考勤方面的得分占总分的 10%。
2. 相互协作完成实验任务方面的得分占总分的 40%。
3. 实验报告方面的得分占总分的 50%。

实验 5 图形、图片与多媒体

5.1 实 验 目 的

在屏幕上绘制各种图形，了解 Android 中的图形、图片处理技术和多媒体技术。

5.2 实 验 要 求

- 了解在屏幕上绘图的方法。
- 了解 Android 中图形、图片处理技术的使用。
- 了解 Android 中多媒体技术的使用。

5.3 实 验 内 容

【练习 5.1】探照灯效果

编写 Android 项目，实现探照灯效果。
- 运行效果如图 E5.1 所示。
- 资源文件布局如图 E5.2 所示。

图 E5.1 运行效果

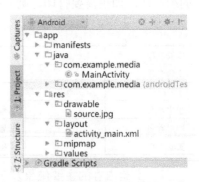

图 E5.2 资源文件布局

- activity_main.xml 文件代码如下：

```xml
<?xml version="1.0" encoding="utf-8"?>
<FrameLayout xmlns:android="http:// schemas.android.com/apk/res/android"
    android:id="@+id/frameLayout1"
    android:layout_width="fill_parent"
    android:layout_height="fill_parent"
    android:orientation="vertical" >
```

```
    </FrameLayout>
```

- **MainActivity.java** 文件代码如下:

```java
package com.example.media;

import android.app.Activity;
import android.content.Context;
import android.graphics.Bitmap;
import android.graphics.BitmapFactory;
import android.graphics.BitmapShader;
import android.graphics.Canvas;
import android.graphics.Matrix;
import android.graphics.Paint;
import android.graphics.Shader.TileMode;
import android.graphics.drawable.ShapeDrawable;
import android.graphics.drawable.shapes.OvalShape;
import android.os.Bundle;
import android.view.MotionEvent;
import android.view.View;
import android.widget.FrameLayout;

public class MainActivity extends Activity {
    @Override
    public void onCreate(Bundle savedInstanceState) {
        super.onCreate(savedInstanceState);
        setContentView(R.layout.activity_main);
        // 获取布局文件中的帧布局管理器
        FrameLayout ll = (FrameLayout) findViewById(R.id.frameLayout1);
        ll.addView(new MyView(this));          // 将自定义视图添加到帧布局管理器中

    }

    public class MyView extends View {
        private Bitmap bitmap;                 // 源图片,也就是背景图片
        private ShapeDrawable drawable;

        private final int RADIUS = 200;        // 探照灯的半径

        private Matrix matrix = new Matrix();

        public MyView(Context context) {
            super(context);
            Bitmap bitmap_source = BitmapFactory.decodeResource(getResources(),
                            R.drawable.source);    // 获取要显示的源图片
            bitmap = bitmap_source;
            BitmapShader shader = new BitmapShader(Bitmap.createScaledBitmap(
                    bitmap_source, bitmap_source.getWidth(),
                    bitmap_source.getHeight(), true), TileMode.CLAMP,
                    TileMode.CLAMP);               // 创建 BitmapShader 对象
```

```
        // 圆形的 drawable 参数
        drawable = new ShapeDrawable(new OvalShape());
        drawable.getPaint().setShader(shader);
        drawable.setBounds(0, 0, RADIUS * 2, RADIUS * 2);// 设置圆的外切矩形
    }

    @Override
    protected void onDraw(Canvas canvas) {
        super.onDraw(canvas);
        Paint p=new Paint();
        p.setAlpha(50);
        canvas.drawBitmap(bitmap, 0,0, p);          // 绘制背景图片
        drawable.draw(canvas);                       // 绘制探照灯照射的图片
    }

    @Override
    public boolean onTouchEvent(MotionEvent event) {
        final int x = (int) event.getX();           // 获取当前触摸点的 X 轴坐标
        final int y = (int) event.getY();           // 获取当前触摸点的 Y 轴坐标
        // 平移到绘制阴影的起始位置
        matrix.setTranslate(RADIUS - x , RADIUS - y );
        drawable.getPaint().getShader().setLocalMatrix(matrix);
        // 设置圆的外切矩形
        drawable.setBounds(x - RADIUS, y - RADIUS, x + RADIUS, y + RADIUS);
        invalidate();                                // 重绘画布
        return true;
    }
}
```

【练习 5.2】实现视频播放器

编写 Android 项目，使用 VideoView 实现视频播放器。
- 运行效果如图 E5.3 所示。
- 资源文件布局如图 E5.4 所示。

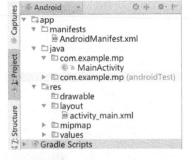

图 E5.3　运行效果　　　　　　　　　　图 E5.4　资源文件布局

- activity_main.xml 文件代码如下：

```xml
<?xml version="1.0" encoding="utf-8"?>
<RelativeLayout xmlns:android="http://schemas.android.com/apk/res/android"
    xmlns:tools="http://schemas.android.com/tools"
    android:layout_width= "match_parent"
    android:layout_height="match_parent"
    android:paddingLeft="@dimen/activity_ horizontal_margin"
    android:paddingRight="@dimen/activity_horizontal_margin"
    android:paddingTop="@dimen/activity_vertical_margin"
    android:paddingBottom="@dimen/activity_vertical_margin"
    tools:context= ".MainActivity">

    <VideoView
        android:id="@+id/VideoView01"
        android:layout_width="fill_parent"
        android:layout_height="fill_parent"
        android:layout_below="@+id/LoadButton" />
    <Button android:id="@+id/LoadButton"
        android:layout_width="wrap_content"
        android:layout_height="wrap_content"
        android:text="装载"
        android:layout_x="30px"
        android:layout_y="300px"
        android:textSize="40dp" />
    <Button android:id="@+id/PlayButton"
        android:layout_width="wrap_content"
        android:layout_height="wrap_content"
        android:text="播放"
        android:layout_x="120px"
        android:layout_y="300px"
        android:textSize="40dp"
        android:layout_alignParentTop="true"
        android:layout_toEndOf="@+id/LoadButton" />
    <Button android:id="@+id/PauseButton"
        android:layout_width="wrap_content"
        android:layout_height="wrap_content"
        android:text="暂停"
        android:layout_x="210px"
        android:layout_y="300px"
        android:textSize="40dp"
        android:layout_alignParentTop="true"
        android:layout_toEndOf="@+id/PlayButton" />
</RelativeLayout>
```

- MainActivity.java 文件代码如下：

```java
package com.example.mp;

import android.app.Activity;
import android.content.pm.ActivityInfo;
import android.os.Bundle;
```

```java
import android.util.Log;
import android.view.View;
import android.view.View.OnClickListener;
import android.view.Window;
import android.view.WindowManager;
import android.widget.Button;
import android.widget.MediaController;
import android.widget.VideoView;

public class MainActivity extends Activity {
    private static final String TAG="VideoView";
    @Override
    protected void onCreate(Bundle savedInstanceState) {
        super.onCreate(savedInstanceState);

        // 不要标题
        requestWindowFeature(Window.FEATURE_NO_TITLE);
        // 设置为全屏模式
        getWindow().setFlags(WindowManager.LayoutParams.FLAG_FULLSCREEN,
                WindowManager.LayoutParams.FLAG_FULLSCREEN);
        // 强制为横屏模式
        setRequestedOrientation(ActivityInfo.SCREEN_ORIENTATION_LANDSCAPE);

        setContentView(R.layout.activity_main);


        final VideoView videoView = (VideoView) findViewById(R.id.VideoView01);


        Button PauseButton = (Button) this.findViewById(R.id.PauseButton);
        Button LoadButton = (Button) this.findViewById(R.id.LoadButton);
        Button PlayButton = (Button) this.findViewById(R.id.PlayButton);


        LoadButton.setOnClickListener(new OnClickListener()
        {
            public void onClick(View arg0)
            {

                videoView.setVideoPath("/mnt/sdcard/apple.mp4");
                videoView.setMediaController(new MediaController(MainActivity.this));
                videoView.requestFocus();
            }
        });

        PlayButton.setOnClickListener(new OnClickListener()
        {
            public void onClick(View arg0)
```

```
            {
                Log.v(TAG, "start");
                videoView.start();
                Log.v(TAG,"start OK");
            }
        });

        PauseButton.setOnClickListener(new OnClickListener()
        {
            public void onClick(View arg0)
            {
                videoView.pause();
            }
        });
    }
}
```

- AndroidManifest.xml 文件代码如下：

```xml
<?xml version="1.0" encoding="utf-8"?>
<manifest xmlns:android="http://schemas.android.com/apk/res/android"
    package="com.example.mp" >

    <application
        android:allowBackup="true"
        android:icon="@mipmap/ic_launcher"
        android:label="@string/app_name"
        android:supportsRtl="true"
        android:theme="@style/AppTheme" >
        <activity android:name=".MainActivity" >
            <intent-filter>
                <action android:name="android.intent.action.MAIN" />

                <category android:name="android.intent.category.LAUNCHER" />
            </intent-filter>
        </activity>
    </application>
    <uses-permission
        android:name="android.permission.WRITE_EXTERNAL_STORAGE"/>
</manifest>
```

- 将 apple.mp4 视频文件导入手机文件系统。

在 Android Studio 菜单栏中选择 "Tools" → "Android" → "Android Device Monitor" 命令，如图 E5.5 所示。

在弹出的 "Android Device Monitor" 界面中，选择 "File Explorer" 选项卡，找到将要导入视频的/mnt/sdcard 目录，然后单击 "导入" 图标，导入 apple.mp4 文件，最终效果如图 E5.6 所示。

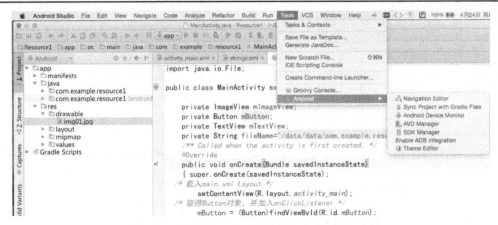

图 E5.5　选择"Android Device Monitor"命令

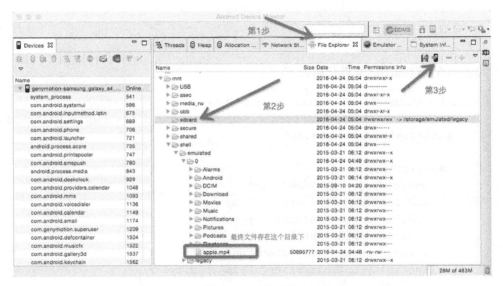

图 E5.6　最终效果

5.4　实 验 报 告

1. 每人一份实验报告，统一使用学校提供的 A4 尺寸的实验报告册书写或者使用 A4 尺寸的纸打印。如果需要打印，则必须采用如表 E5.1 所示的表头格式。

表 E5.1　表头格式

实验课程						
实验名称						
实验时间	学年	学期	周 星期	．	第　节	
学生姓名		学号			班级	
同组姓名		学号			班级	
实验地点		设备号			指导教师	

2. 实验内容的主要结果及对结果的分析。

3. 实验过程中所遇到的问题的解决办法。
4. 心得体会、意见和建议。

5.5 实验成绩考核

1. 考勤方面的得分占总分的 10%。
2. 相互协作完成实验任务方面的得分占总分的 40%。
3. 实验报告方面的得分占总分的 50%。

实验 6 Android 网络编程基础

6.1 实 验 目 的

本次实验的目的是让大家熟悉 Android 开发中如何获取天气预报，包括了解和熟悉 WebView 的使用、网络编程、事件处理等内容。

6.2 实 验 要 求

- 了解和熟悉 WebView 的使用。
- 了解 Android 的网络编程。

6.3 实 验 内 容

【练习 6.1】基于 WebView 获取天气预报

- 运行效果如图 E6.1 所示。
- 资源文件布局如图 E6.2 所示。

图 E6.1 运行效果   图 E6.2 资源文件布局

- activity_main.xml 文件代码如下：

```xml
<?xml version="1.0" encoding="utf-8"?>
<LinearLayout xmlns:android="http://schemas.android.com/apk/res/android"
    android:orientation="vertical"
    android:gravity="center_horizontal"
    android:layout_width="fill_parent"
    android:layout_height="fill_parent">
    <LinearLayout
        android:orientation="horizontal"
        android:layout_width="wrap_content"
        android:layout_height="wrap_content">

        <Button
            android:id="@+id/bj"
            android:layout_width="wrap_content"
            android:layout_height="wrap_content"
            android:text="@string/bj"
            android:textSize="30dp" />

        <Button
            android:id="@+id/sh"
            android:layout_width="wrap_content"
            android:layout_height="wrap_content"
            android:text="@string/sh"
            android:textSize="30dp" />

        <Button
            android:id="@+id/heb"
            android:layout_width="wrap_content"
            android:layout_height="wrap_content"
            android:text="@string/heb"
            android:textSize="30dp" />

    </LinearLayout>
    <LinearLayout
        android:orientation="horizontal"
        android:layout_width="wrap_content"
        android:layout_height="wrap_content">
    <Button
        android:id="@+id/gz"
        android:layout_width="wrap_content"
        android:layout_height="wrap_content"
        android:text="@string/gz"
        android:textSize="30dp" />

    <Button
        android:id="@+id/cc"
        android:layout_width="wrap_content"
```

```xml
            android:layout_height="wrap_content"
            android:text="@string/cc"
            android:textSize="30dp" />

        <Button
            android:id="@+id/sy"
            android:layout_width="wrap_content"
            android:layout_height="wrap_content"
            android:text="@string/sy"
            android:textSize="30dp"
            android:layout_gravity="right" />
    </LinearLayout>
    <WebView android:id="@+id/webView1"
        android:layout_width="match_parent"
        android:layout_height="0dip"
        android:focusable="false"
        android:layout_weight="1" />

</LinearLayout>
```

- strings.xml 文件代码如下：

```xml
<resources>
    <string name="app_name">WebView</string>
    <string name="go">GO</string>
    <string name="bj">北京</string>
    <string name="sh">上海</string>
    <string name="gz">广州</string>
    <string name="heb">哈尔滨</string>
    <string name="cc">长春</string>
    <string name="sy">沈阳</string>
</resources>
```

- MainActivity.java 文件代码如下：

```java
package com.example.webview;

import android.app.Activity;
import android.os.Bundle;
import android.view.View;
import android.view.View.OnClickListener;
import android.webkit.WebChromeClient;
import android.webkit.WebView;
import android.webkit.WebViewClient;
import android.widget.Button;

public class MainActivity extends Activity implements OnClickListener {
    private WebView webView;     // 声明 WebView 对象
```

```java
@Override
protected void onCreate(Bundle savedInstanceState) {
    super.onCreate(savedInstanceState);
    setContentView(R.layout.activity_main);
    // 获取 WebView
    webView=(WebView)findViewById(R.id.webView1);
    // 设置 JavaScript 可用
    webView.getSettings().setJavaScriptEnabled(true);
    // 处理 JavaScript 对话框
    webView.setWebChromeClient(new WebChromeClient());
    // 处理各种通知和请求事件。如果不使用这条代码，将使用内置浏览器访问网页
    webView.setWebViewClient(new WebViewClient());
    // 设置默认显示的天气预报信息
    webView.loadUrl("http://m.weather.com.cn/mweather/");
    webView.setInitialScale(57*4);    // 将网页内容放大 4 倍
    // 获取布局管理器中添加的"北京"按钮
    Button bj=(Button)findViewById(R.id.bj);
    bj.setOnClickListener(this);
    // 获取布局管理器中添加的"上海"按钮
    Button sh=(Button)findViewById(R.id.sh);
    sh.setOnClickListener(this);
    // 获取布局管理器中添加的"哈尔滨"按钮
    Button heb=(Button)findViewById(R.id.heb);
    heb.setOnClickListener(this);
    // 获取布局管理器中添加的"长春"按钮
    Button cc=(Button)findViewById(R.id.cc);
    cc.setOnClickListener(this);
    // 获取布局管理器中添加的"沈阳"按钮
    Button sy=(Button)findViewById(R.id.sy);
    sy.setOnClickListener(this);
    // 获取布局管理器中添加的"广州"按钮
    Button gz=(Button)findViewById(R.id.gz);
    gz.setOnClickListener(this);
}
@Override
public void onClick(View view){
    switch(view.getId()){
        case R.id.bj:      // 单击的是"北京"按钮
            openUrl("101010100");
            break;
        case R.id.sh:      // 单击的是"上海"按钮
            openUrl("101020100");
            break;
        case R.id.heb:     // 单击的是"哈尔滨"按钮
            openUrl("101050101");
            break;
        case R.id.cc:      // 单击的是"长春"按钮
```

```
            openUrl("101060101");
            break;
        case R.id.sy:      // 单击的是"沈阳"按钮
            openUrl("101070101");
            break;
        case R.id.gz:      // 单击的是"广州"按钮
            openUrl("101280101");
            break;
        }
    }
    // 打开网页的方法
    private void openUrl(String id){
        // 获取并显示天气预报信息
        webView.loadUrl("http://m.weather.com.cn/mweather/"+ id +".shtml" );
    }
}
```

- AndroidManifest.xml 文件代码如下:

```
<?xml version="1.0" encoding="utf-8"?>
<manifest xmlns:android="http://schemas.android.com/apk/res/android"
    package="com.example.webview" >
    <uses-permission android:name="android.permission.INTERNET"/>
    <application
        android:allowBackup="true"
        android:icon="@mipmap/ic_launcher"
        android:label="@string/app_name"
        android:supportsRtl="true"
        android:theme="@style/AppTheme" >
        <activity android:name=".MainActivity" >
            <intent-filter>
                <action android:name="android.intent.action.MAIN" />

                <category android:name="android.intent.category.LAUNCHER" />
            </intent-filter>
        </activity>
    </application>

</manifest>
```

【练习 6.2】基于 WebService 的手机归属地查询
- 运行效果如图 E6.3 所示。
- 将 ksoap2-android 的 JAR 包添加到 Android 工程项目中,步骤如下。

首先,在 Google 提供的项目下载网站中下载开发包,网址为"https://code.google.com/archive/p/ksoap2-android/source",单击"Downloads"按钮,选择 ksoap2-android-assembly-2.4-jar-with-dependencies.jar,如图 E6.4 所示。

图 E6.3　运行效果　　　　　　　　　图 E6.4　下载开发包

也可以直接在本书资源中找到该 JAR 包，并进行下载和安装。

然后，将下载的 ksoap2-android 的 JAR 包添加到工程项目的 libs 目录下，并右击该包，在弹出的快捷菜单中选择"Add as library"命令，这样就将 ksoap2-android 集成到 Android 项目中了，如图 E6.5 所示。

- 根据网上开放的 WebService 服务开发自己的应用程序，本节将展示如何实现一个手机归属地查询的小应用，我们使用的 WebService 服务地址为 http://ws.webxml.com.cn/WebServices/MobileCodeWS.asmx。

在打开该网址后，单击"getMobileCodeInfo"链接，进入 WebService 服务页，如图 E6.6 所示。

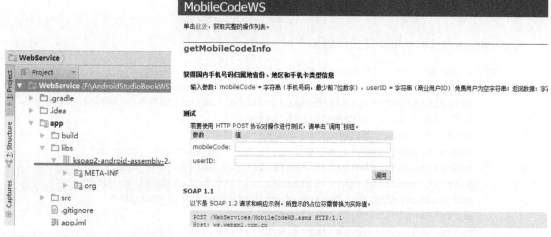

图 E6.5　集成 ksoap2-android　　　　　　图 E6.6　WebService 服务页

在 http://ws.webxml.com.cn/WebServices/MobileCodeWS.asmx 后面加上"?wsdl"就可以访问其 wsdl 说明页了，如图 E6.7 所示。

```
▼<wsdl:definitions xmlns:soap="http://schemas.xmlsoap.org/wsdl/soap/" xmlns:tm="http://microsoft.com/wsdl/mime/textMatching/"
  xmlns:soapenc="http://schemas.xmlsoap.org/soap/encoding/" xmlns:mime="http://schemas.xmlsoap.org/wsdl/mime/" xmlns:tns="http://WebXml.com.cn/"
  xmlns:s="http://www.w3.org/2001/XMLSchema" xmlns:soap12="http://schemas.xmlsoap.org/wsdl/soap12/" xmlns:http="http://schemas.xmlsoap.org/wsdl/http/"
  xmlns:wsdl="http://schemas.xmlsoap.org/wsdl/" targetNamespace="http://WebXml.com.cn/">
  ▼<wsdl:documentation xmlns:wsdl="http://schemas.xmlsoap.org/wsdl/">
    <a href="http://www.webxml.com.cn/" target="_blank">WebXml.com.cn</a> <strong>国内手机号码归属地查询WEB服务</strong>，提供最新的国内手机号码段归属
    用本站 WEB 服务请注明或链接本站：<a href="http://www.webxml.com.cn/" target="_blank">http://www.webxml.com.cn/</a> 感谢大家的支持！<br /> 
  </wsdl:documentation>
  ▼<wsdl:types>
    ▼<s:schema elementFormDefault="qualified" targetNamespace="http://WebXml.com.cn/">
      ▼<s:element name="getMobileCodeInfo">
        ▼<s:complexType>
          ▼<s:sequence>
            <s:element minOccurs="0" maxOccurs="1" name="mobileCode" type="s:string"/>
            <s:element minOccurs="0" maxOccurs="1" name="userID" type="s:string"/>
          </s:sequence>
        </s:complexType>
      </s:element>
      ▼<s:element name="getMobileCodeInfoResponse">
        ▼<s:complexType>
          ▼<s:sequence>
            <s:element minOccurs="0" maxOccurs="1" name="getMobileCodeInfoResult" type="s:string"/>
          </s:sequence>
```

图 E6.7 wsdl 说明页

从图 E6.7 中，我们可以得到几个很关键的点。
① 作用域 targetNamespace = http://WebXml.com.cn/。
② 查询的方法名为"getMobileCodeInfo"，需要带上"mobileCode"与"userID"两个参数。
③ 返回的结果被保存在"getMobileCodeInfoResult"中。
- 资源文件布局如图 E6.8 所示。

图 E6.8 资源文件布局

- activity_web_client.xml 文件代码如下：

```xml
<?xml version="1.0" encoding="utf-8"?>
<LinearLayout xmlns:android="http://schemas.android.com/apk/res/android"
    xmlns:tools="http://schemas.android.com/tools"
    android:layout_width="match_parent"
    android:layout_height="match_parent"
    android:paddingBottom="@dimen/activity_vertical_margin"
    android:paddingLeft="@dimen/activity_horizontal_margin"
    android:paddingRight="@dimen/activity_horizontal_margin"
    android:paddingTop="@dimen/activity_vertical_margin"
    android:orientation="vertical"
    tools:context="com.example.webservice.WebClient">

    <LinearLayout
        android:layout_width="match_parent"
        android:layout_height="wrap_content"
        android:orientation="horizontal">
        <TextView
```

```xml
            android:layout_width="wrap_content"
            android:layout_height="wrap_content"
            android:text="输入手机号："/>

        <EditText
            android:layout_width="150dp"
            android:layout_height="wrap_content"
            android:id="@+id/etphone" />

        <Button
            android:layout_width="wrap_content"
            android:layout_height="wrap_content"
            android:text="搜索"
            android:id="@+id/btnsearch" />

    </LinearLayout>

    <TextView
        android:layout_width="wrap_content"
        android:layout_height="wrap_content"
        android:text="查询结果："/>

    <TextView
        android:id="@+id/tvinfo"
        android:layout_width="wrap_content"
        android:layout_height="wrap_content"/>

</LinearLayout>
```

- WebClient.java 文件代码如下：

```java
package com.example.webservice;

import android.os.AsyncTask;
import android.support.v7.app.AppCompatActivity;
import android.os.Bundle;
import android.view.View;
import android.widget.Button;
import android.widget.EditText;
import android.widget.TextView;

import org.ksoap2.SoapEnvelope;
import org.ksoap2.SoapFault;
import org.ksoap2.serialization.SoapObject;
import org.ksoap2.serialization.SoapSerializationEnvelope;
import org.ksoap2.transport.HttpTransportSE;
import org.xmlpull.v1.XmlPullParserException;

import java.io.IOException;

public class WebClient extends AppCompatActivity {
```

```java
            private static final String SERVER_URL = "http://ws.webxml.com.cn/
WebServices/MobileCodeWS.asmx?wsdl";
            // 调用的 WebService 命名空间
            private static final String PACE = "http://WebXml.com.cn/";
            // 获取归属地的方法名
            private static final String W_NAME = "getMobileCodeInfo";

            private EditText etPhone;
            private Button btnSearch;
            private TextView tvInfo;

            @Override
            protected void onCreate(Bundle savedInstanceState) {
                super.onCreate(savedInstanceState);
                setContentView(R.layout.activity_web_client);
                etPhone = (EditText) findViewById(R.id.etphone);
                btnSearch = (Button) findViewById(R.id.btnsearch);
                tvInfo = (TextView) findViewById(R.id.tvinfo);

                btnSearch.setOnClickListener(new View.OnClickListener() {
                    @Override
                    public void onClick(View v) {
                        String cityName = etPhone.getText().toString();
                        if (cityName.length() > 0) {
                            getWeatherInfo(etPhone.getText().toString());
                        }
                    }
                });
            }

            private void getWeatherInfo(String phoneMum){
                new AsyncTask<String, Void, String>() {
                    @Override
                    protected String doInBackground(String... params) {
                        String local = "";
                        final HttpTransportSE httpSe = new HttpTransportSE(SERVER_URL);
                        httpSe.debug = true;
                        SoapObject soapObject = new SoapObject(PACE, W_NAME);
                        soapObject.addProperty("mobileCode", params[0]);
                        soapObject.addProperty("userID", "");
                        final SoapSerializationEnvelope serializa=new SoapSerializationEnvelope(
                                SoapEnvelope.VER10);
                        serializa.setOutputSoapObject(soapObject);
                        serializa.dotNet = true;
                        // 获取返回信息
                        try {
                            httpSe.call(PACE + W_NAME, serializa);
                            if (serializa.getResponse() != null) {
                                SoapObject result = (SoapObject) serializa.bodyIn;
                                local = result.getProperty("getMobileCodeInfoResult").
toString();
```

```
                }
            }
            catch (XmlPullParserException e) {
                e.printStackTrace();
            }catch (SoapFault soapFault) {
                soapFault.printStackTrace();
            }catch (IOException e) {
                e.printStackTrace();
            }
            return local;
        }

        @Override
        protected void onPostExecute(String result) {
            tvInfo.setText(result);
        }
    }.execute(phoneMum);
    }
}
```

- **AndroidManifest.xml 文件代码如下：**

```
<?xml version="1.0" encoding="utf-8"?>
<manifest xmlns:android="http://schemas.android.com/apk/res/android"
    package="com.example.webservice">

    <uses-permission android:name="android.permission.INTERNET"/>

    <application
        android:allowBackup="true"
        android:icon="@mipmap/ic_launcher"
        android:label="@string/app_name"
        android:supportsRtl="true"
        android:theme="@style/AppTheme">
        <activity android:name="com.example.webservice.WebClient">
            <intent-filter>
                <action android:name="android.intent.action.MAIN" />

                <category android:name="android.intent.category.LAUNCHER" />
            </intent-filter>
        </activity>
    </application>

</manifest>
```

6.4 实 验 报 告

1. 每人一份实验报告，统一使用学校提供的 A4 尺寸的实验报告册书写或者使用 A4 尺寸的纸打印。如果需要打印，则必须采用如表 E6.1 所示的表头格式。

表 E6.1　表头格式

实验课程									
实验名称									
实验时间		学年	学期	周	星期	第　节			
学生姓名			学号				班级		
同组姓名			学号				班级		
实验地点			设备号				指导教师		

2. 实验内容的主要结果及对结果的分析。
3. 实验过程中所遇到的问题的解决办法。
4. 心得体会、意见和建议。

6.5　实验成绩考核

1. 考勤方面的得分占总分的 10%。
2. 相互协作完成实验任务方面的得分占总分的 40%。
3. 实验报告方面的得分占总分的 50%。

实验 7　SQLite 和 SQLiteDatabase 的使用

7.1　实　验　目　的

本次实验的目的是让大家熟悉 Android 中对数据库进行操作的相关接口、类等。SQLiteDatabase 是在 Android 中进行数据库操作时使用最频繁的一个类。通过它可以实现创建或打开数据库、创建数据表、插入数据、删除数据、查询数据、修改数据等操作。

7.2　实　验　要　求

- 实现便签管理小程序。
- 创建项目并熟悉文件目录结构。
- 实现便签的"增删改查"功能的实验步骤。

7.3　实　验　内　容

【练习 7.1】　便签管理小程序
- 程序运行效果如图 E7.1 所示。

activity_main.xml（启动窗体）

insertinfo.xml（新增便签窗体）

showinfo.xml（查看便签信息窗体）

manageflag.xml（便签管理窗体）

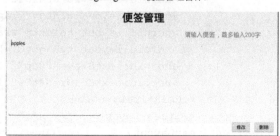

图 E7.1　程序运行效果

- 资源文件布局如图 E7.2 所示。

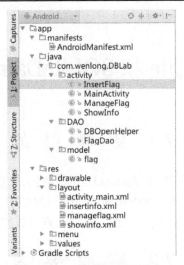

图 E7.2 资源文件布局

- activity_main.xml（启动窗体）文件代码如下：

```
<LinearLayout xmlns:android="http://schemas.android.com/apk/res/android"
    android:layout_width="fill_parent"
    android:layout_height="fill_parent"
    android:orientation="vertical" >
<LinearLayout android:id="@+id/linearLayout1"
    android:layout_height="wrap_content"
    android:layout_width="match_parent"
    android:orientation="vertical"
    android:layout_weight="0.06">
    <RelativeLayout android:layout_height="wrap_content"
    android:layout_width="match_parent">
    <Button android:text="便签信息"
        android:id="@+id/btnflaginfo"
        android:layout_width="wrap_content"
        android:layout_height="wrap_content"
        android:textSize="20dp"
        android:textColor="#8C6931"/>
    <Button android:text="添加便签"
        android:id="@+id/btninsertinfo"
        android:layout_width="wrap_content"
        android:layout_height="wrap_content"
        android:layout_toRightOf="@id/btnflaginfo"
        android:textSize="20dp"
        android:textColor="#8C6931"/>
    </RelativeLayout>
    </LinearLayout>
</LinearLayout>
```

- insertinfo.xml（新增便签窗体）文件代码如下：

```
<?xml version="1.0" encoding="utf-8"?>
```

```xml
<LinearLayout xmlns:android="http://schemas.android.com/apk/res/android"
   android:id="@+id/itemflag"
   android:orientation="vertical"
   android:layout_width="fill_parent"
   android:layout_height="fill_parent">
   <LinearLayout
      android:orientation="vertical"
      android:layout_width="fill_parent"
      android:layout_height="fill_parent"
      android:layout_weight="3">
      <TextView
         android:layout_width="wrap_content"
         android:layout_gravity="center"
         android:gravity="center_horizontal"
         android:text="新增便签"
         android:textSize="40sp"
         android:textColor="#000000"
         android:textStyle="bold"
         android:layout_height="wrap_content"/>
   </LinearLayout>
   <LinearLayout
      android:orientation="vertical"
      android:layout_width="fill_parent"
      android:layout_height="fill_parent"
      android:layout_weight="1">
      <RelativeLayout android:layout_width="fill_parent"
          android:layout_height="fill_parent"
          android:padding="5dp">
          <TextView android:layout_width="350dp"
             android:id="@+id/tvFlag"
             android:textSize="23sp"
             android:text="请输入便签，最多输入200字"
             android:textColor="#8C6931"
             android:layout_alignParentRight="true"
             android:layout_height="wrap_content" />
          <EditText
             android:id="@+id/txtFlag"
             android:layout_width="350dp"
             android:layout_height="400dp"
             android:layout_below="@id/tvFlag"
             android:gravity="top"
             android:singleLine="false"/>
      </RelativeLayout>
   </LinearLayout>
   <LinearLayout
      android:orientation="vertical"
      android:layout_width="fill_parent"
      android:layout_height="fill_parent"
      android:layout_weight="3">
```

```xml
            <RelativeLayout android:layout_width="fill_parent"
                android:layout_height="fill_parent"
                android:padding="10dp">
            <Button
                android:id="@+id/btnflagCancel"
                android:layout_width="80dp"
                android:layout_height="wrap_content"
                android:layout_alignParentRight="true"
                android:layout_marginLeft="10dp"
                android:text="取消"/>
            <Button
                android:id="@+id/btnflagSave"
                android:layout_width="80dp"
                android:layout_height="wrap_content"
                android:layout_toLeftOf="@id/btnflagCancel"
                android:text="保存"
                android:maxLength="200"/>
        </RelativeLayout>
    </LinearLayout>
</LinearLayout>
```

- **showinfo.xml（查看便签信息窗体）文件代码如下：**

```xml
<?xml version="1.0" encoding="utf-8"?>
<LinearLayout xmlns:android="http://schemas.android.com/apk/res/android"
    android:layout_width="match_parent"
    android:layout_height="match_parent"
    android:orientation="vertical" >

    <TextView
        android:id="@+id/textView1"
        android:layout_width="wrap_content"
        android:layout_height="wrap_content"
        android:textSize="20dp"
        android:text="便签信息" />
    <LinearLayout android:id="@+id/linearLayout2"
        android:layout_height="wrap_content"
        android:layout_width="match_parent"
        android:orientation="vertical"
        android:layout_weight="0.94">
        <ListView android:id="@+id/lvinfo"
            android:layout_width="match_parent"
            android:layout_height="match_parent"
            android:scrollbarAlwaysDrawVerticalTrack="true"/>
    </LinearLayout>
</LinearLayout>
```

- **manageflag.xml（便签管理窗体）文件代码如下：**

```xml
<?xml version="1.0" encoding="utf-8"?>
<LinearLayout xmlns:android="http://schemas.android.com/apk/res/android"
```

```xml
    android:id="@+id/flagmanage"
    android:orientation="vertical"
    android:layout_width="fill_parent"
    android:layout_height="fill_parent">
    <LinearLayout
        android:orientation="vertical"
        android:layout_width="fill_parent"
        android:layout_height="fill_parent"
        android:layout_weight="3">
        <TextView
            android:layout_width="wrap_content"
            android:layout_gravity="center"
            android:gravity="center_horizontal"
            android:text="便签管理"
            android:textSize="40sp"
            android:textColor="#000000"
            android:textStyle="bold"
            android:layout_height="wrap_content"/>
    </LinearLayout>
    <LinearLayout
        android:orientation="vertical"
        android:layout_width="fill_parent"
        android:layout_height="fill_parent"
        android:layout_weight="1">
        <RelativeLayout android:layout_width="fill_parent"
            android:layout_height="fill_parent"
            android:padding="5dp">
            <TextView android:layout_width="350dp"
            android:id="@+id/tvFlagManage"
            android:textSize="23sp"
            android:text="请输入便签,最多输入200字"
            android:textColor="#8C6931"
            android:layout_alignParentRight="true"
            android:layout_height="wrap_content"/>
          <EditText
            android:id="@+id/txtFlagManage"
            android:layout_width="350dp"
            android:layout_height="400dp"
            android:layout_below="@id/tvFlagManage"
            android:gravity="top"
            android:singleLine="false"/>
        </RelativeLayout>
    </LinearLayout>
    <LinearLayout
        android:orientation="vertical"
        android:layout_width="fill_parent"
        android:layout_height="fill_parent"
        android:layout_weight="3">
        <RelativeLayout android:layout_width="fill_parent"
```

```xml
        android:layout_height="fill_parent"
        android:padding="10dp">
    <Button
        android:id="@+id/btnFlagManageDelete"
        android:layout_width="80dp"
        android:layout_height="wrap_content"
        android:layout_alignParentRight="true"
        android:layout_marginLeft="10dp"
        android:text="删除"/>
    <Button
        android:id="@+id/btnFlagManageEdit"
        android:layout_width="80dp"
        android:layout_height="wrap_content"
        android:layout_toLeftOf="@id/btnFlagManageDelete"
        android:text="修改"
        android:maxLength="200"/>
    </RelativeLayout>
  </LinearLayout>
</LinearLayout>
```

- **MainActivity.java 文件代码如下：**

```java
package com.wenlong.DBLab.activity;

import android.os.Bundle;
import android.app.Activity;
import android.content.Intent;
import android.view.Menu;
import android.view.View;
import android.view.View.OnClickListener;
import android.widget.Button;

public class MainActivity extends Activity {
Button btnflaginfo,btninsertinfo;
    @Override
    protected void onCreate(Bundle savedInstanceState) {
        super.onCreate(savedInstanceState);
        setContentView(R.layout.activity_main);
        btnflaginfo=(Button)findViewById(R.id.btnflaginfo);
        btninsertinfo=(Button)findViewById(R.id.btninsertinfo);

        btnflaginfo.setOnClickListener(new OnClickListener() {

            @Override
            public void onClick(View v) {
                Intent intent=new Intent(MainActivity.this,ShowInfo.class);
                startActivity(intent);

            }
        });
```

实验7 SQLite 和 SQLiteDatabase 的使用

```java
        btninsertinfo.setOnClickListener(new OnClickListener() {

            @Override
            public void onClick(View v) {
                Intent intent=new Intent(MainActivity.this,InsertFlag.class);
                startActivity(intent);


            }
        });
    }

    @Override
    public boolean onCreateOptionsMenu(Menu menu) {
        // Inflate the menu; this adds items to the action bar if it is present.
        getMenuInflater().inflate(R.menu.main, menu);
        return true;
    }


}
```

- **InsertFlag.java 文件代码如下:**

```java
package com.wenlong.DBLab.activity;

import android.app.Activity;
import android.os.Bundle;
import android.view.View;
import android.view.View.OnClickListener;
import android.widget.Button;
import android.widget.EditText;
import android.widget.Toast;

import com.wenlong.DBLab.DAO.FlagDao;
import com.wenlong.DBLab.model.flag;

public class InsertFlag extends Activity {
    EditText txtFlag;                           // 创建 EditText 对象
    Button btnflagSaveButton;                   // 创建 Button 对象
    Button btnflagCancelButton;                 // 创建 Button 对象

    @Override
    protected void onCreate(Bundle savedInstanceState) {
        super.onCreate(savedInstanceState);
        setContentView(R.layout.insertinfo);
        txtFlag=(EditText)findViewById(R.id.txtFlag);
        btnflagSaveButton=(Button)findViewById(R.id.btnflagSave);
        btnflagCancelButton=(Button)findViewById(R.id.btnflagCancel);
        btnflagSaveButton.setOnClickListener(new OnClickListener() {

            @Override
            public void onClick(View v) {
```

```java
            String strFlag = txtFlag.getText().toString();   // 获取便签文本框的值
            if (!strFlag.isEmpty()) {                        // 判断获取
                FlagDao flagDAO = new FlagDao(InsertFlag.this);  // 创建 FlagDAO 对象
                flag flag = new flag(
                    flagDAO.getMaxId() + 1, strFlag);        // 创建 Tb_flag 对象
                flagDAO.add(flag);                           // 添加便签信息
                // 弹出信息提示
                Toast.makeText(InsertFlag.this, "〖新增便签〗数据添加成功！",
                    Toast.LENGTH_SHORT).show();
            } else {
                Toast.makeText(InsertFlag.this, "请输入便签！",
                    Toast.LENGTH_SHORT).show();
            }
        }
    });
    btnflagCancelButton.setOnClickListener(new OnClickListener() {
        @Override
        public void onClick(View v) {
            finish();
        }
    });
  }
}
```

- ShowInfo.java 文件代码如下：

```java
package com.wenlong.DBLab.activity;

import android.app.Activity;
import android.content.Intent;
import android.os.Bundle;
import android.view.View;
import android.widget.AdapterView;
import android.widget.AdapterView.OnItemClickListener;
import android.widget.ArrayAdapter;
import android.widget.ListView;
import android.widget.TextView;

import com.wenlong.DBLab.DAO.FlagDao;
import com.wenlong.DBLab.model.flag;

import java.util.List;

public class ShowInfo extends Activity {
    public static final String FLAG = "id";    // 定义一个常量，用来作为请求码
    ListView lvinfo;                            // 创建 ListView 对象
    String[] strInfos = null;                   // 定义字符串数组，用来存储收入信息
    ArrayAdapter<String> arrayAdapter = null;   // 创建 ArrayAdapter 对象
    @Override
    protected void onCreate(Bundle savedInstanceState) {
        super.onCreate(savedInstanceState);
```

```
setContentView(R.layout.showinfo);
lvinfo=(ListView)findViewById(R.id.lvinfo);
FlagDao flaginfo = new FlagDao(ShowInfo.this);    // 创建 FlagDAO 对象
// 获取所有便签信息, 并将它们存储到 List 泛型集合中
List<flag> listFlags = flaginfo.getScrollData(0,
        (int) flaginfo.getCount());
strInfos = new String[listFlags.size()];// 设置字符串数组的长度
int n = 0;                                 // 定义一个开始标识
for (flag tb_flag : listFlags)
{
    // 将便签相关信息组合成一个字符串, 并将其存储到字符串数组的相应位置
    strInfos[n] = tb_flag.getid() + "|" + tb_flag.getFlag();
    if (strInfos[n].length() > 15)          //判断便签信息的长度是否大于15
        // 将索引值大于15的位置后面的字符串用 "……" 代替
        strInfos[n] = strInfos[n].substring(0, 15) + "……";
    n++;                                    // 标识加1
}
arrayAdapter = new ArrayAdapter<String>(this,
        android.R.layout.simple_list_item_1, strInfos);
lvinfo.setAdapter(arrayAdapter);
lvinfo.setOnItemClickListener(new OnItemClickListener() {
    @Override
    public void onItemClick(AdapterView<?> parent, View view, int position,
            long id) {
        // 记录单击的项信息
        String strInfo = String.valueOf(((TextView) view).getText());
        // 从项信息中截取编号
        String strid = strInfo.substring(0, strInfo.indexOf('|'));
        Intent intent = null;                // 创建 Intent 对象
        // 使用 FlagManage 窗口初始化 Intent 对象
        intent = new Intent(ShowInfo.this, ManageFlag.class);
        intent.putExtra(FLAG, strid);         // 设置要传递的数据
        startActivity(intent);                // 执行 Intent, 打开相应的 Activity
    }
});
    }
}
```

- **ManageFlag.java 文件代码如下:**

```
package com.wenlong.DBLab.activity;

import android.app.Activity;
import android.content.Intent;
import android.os.Bundle;
import android.view.View;
import android.view.View.OnClickListener;
import android.widget.Button;
import android.widget.EditText;
import android.widget.Toast;
```

```java
import com.wenlong.DBLab.DAO.FlagDao;
import com.wenlong.DBLab.model.flag;
public class ManageFlag extends Activity {
    EditText txtFlag;                              // 创建 EditText 对象
    Button btnEdit, btnDel;                        // 创建两个 Button 对象
    String strid;                                  // 创建字符串,用于表示便签的 ID
    @Override
    protected void onCreate(Bundle savedInstanceState) {
        super.onCreate(savedInstanceState);
        setContentView(R.layout.manageflag);
        txtFlag=(EditText)findViewById(R.id.txtFlagManage);
        btnEdit=(Button)findViewById(R.id.btnFlagManageEdit);
        btnDel=(Button)findViewById(R.id.btnFlagManageDelete);
        Intent intent = getIntent();               // 创建 Intent 对象
        Bundle bundle = intent.getExtras();        // 获取便签的 ID
        strid = bundle.getString(ShowInfo.FLAG);   // 将便签的 ID 转换为字符串
        final FlagDao flagDAO = new FlagDao(ManageFlag.this); // 创建 FlagDAO 对象
        txtFlag.setText(flagDAO.find(Integer.parseInt(strid)).getFlag());
        // 为"修改"按钮设置监听事件
        btnEdit.setOnClickListener(new OnClickListener() {
            @Override
            public void onClick(View v) {
                flag tb_flag = new flag();                    // 创建 Tb_flag 对象
                tb_flag.setid(Integer.parseInt(strid));       // 设置便签的 ID
                tb_flag.setFlag(txtFlag.getText().toString()); // 设置便签值
                flagDAO.update(tb_flag);                      // 修改便签信息
                // 弹出信息提示
                Toast.makeText(ManageFlag.this, "〖便签数据〗修改成功!",
                    Toast.LENGTH_SHORT).show();

            }
        });
        // 为"删除"按钮设置监听事件
        btnDel.setOnClickListener(new OnClickListener() {

            @Override
            public void onClick(View v) {
                flagDAO.delete(Integer.parseInt(strid)); // 根据指定的 ID 删除便签信息
                Toast.makeText(ManageFlag.this, "〖便签数据〗删除成功!",
                    Toast.LENGTH_SHORT).show();
            }
        });
    }
}
```

- DBOpenHelper.java 文件代码如下:

```java
package com.wenlong.DBLab.DAO;
```

```java
import android.content.Context;
import android.database.sqlite.SQLiteDatabase;
import android.database.sqlite.SQLiteOpenHelper;

public class DBOpenHelper extends SQLiteOpenHelper {

    private static final int VERSION = 1;              // 定义数据库版本号
    private static final String DBNAME = "flag.db";    // 定义数据库名
    public DBOpenHelper(Context context) {
        super(context, DBNAME, null, VERSION);
    }
    @Override
    public void onCreate(SQLiteDatabase db)             // 创建数据库
    {
        db.execSQL("create table tb_flag (_id integer primary key,flag varchar(200))");
                                                        // 创建便签信息表
    }
    @Override
    // 覆写基类的 onUpgrade()方法，以便更新数据库版本
    public void onUpgrade(SQLiteDatabase db, int oldVersion, int newVersion)
    {
    }
}
```

- FlagDao.java 文件代码如下：

```java
package com.wenlong.DBLab.DAO;
import java.util.ArrayList;
import java.util.List;
import android.content.Context;
import android.database.Cursor;
import android.database.sqlite.SQLiteDatabase;
import com.wenlong.DBLab.DAO.DBOpenHelper;
import com.wenlong.DBLab.model.flag;
public class FlagDao {
    private DBOpenHelper helper;          // 创建 DBOpenHelper 对象
    private SQLiteDatabase db;            // 创建 SQLiteDatabase 对象

    public FlagDao(Context context)       // 定义构造函数
    {
        helper = new DBOpenHelper(context);   // 初始化 DBOpenHelper 对象
    }
    /**
     * 添加便签信息
     *
     * @param tb_flag
     */
    public void add(flag flag) {
        db = helper.getWritableDatabase();           // 初始化 SQLiteDatabase 对象
        db.execSQL("insert into tb_flag (_id,flag) values (?,?)", new Object[] {
```

```java
                    flag.getid(), flag.getFlag() });      // 执行添加便签信息操作
    }
    /**
     * 更新便签信息
     *
     * @param tb_flag
     */
    public void update(flag tb_flag) {
        db = helper.getWritableDatabase();            // 初始化 SQLiteDatabase 对象
        db.execSQL("update tb_flag set flag = ? where _id = ?", new Object[] {
                tb_flag.getFlag(), tb_flag.getid() });    // 执行修改便签信息操作
    }
    /**
     * 查找便签信息
     *
     * @param id
     * @return
     */
    public flag find(int id) {
        db = helper.getWritableDatabase();            // 初始化 SQLiteDatabase 对象
        Cursor cursor = db.rawQuery(
                "select _id,flag from tb_flag where _id = ?",
                // 根据编号查找便签信息，并存储到 Cursor 类中
                new String[] { String.valueOf(id) });
        if (cursor.moveToNext())                      // 遍历找到的便签信息
        {
            // 将遍历到的便签信息存储到 Tb_flag 类中
            return new flag(cursor.getInt(cursor.getColumnIndex("_id")),
                    cursor.getString(cursor.getColumnIndex("flag")));
        }
        return null;                                  // 如果没有信息，则返回 null
    }
    /**
     * 删除便签信息
     *
     * @param ids
     */
    public void delete(Integer... ids) {
        if (ids.length > 0)                           // 判断是否存在要删除的 ID
        {
            StringBuffer sb = new StringBuffer();// 创建 StringBuffer 对象
            for (int i = 0; i < ids.length; i++)  // 遍历要删除的 ID 集合
            {
                sb.append('?').append(',');  // 将删除条件添加到 StringBuffer 对象中
            }
            sb.deleteCharAt(sb.length() - 1);     // 去掉最后一个 "," 字符
            db = helper.getWritableDatabase();    // 创建 SQLiteDatabase 对象
            // 执行删除便签信息操作
            db.execSQL("delete from tb_flag where _id in (" + sb + ")",
```

```java
                (Object[]) ids);
    }
}

/**
 * 获取便签信息
 *
 * @param start
 *            起始位置
 * @param count
 *            每页显示数量
 * @return
 */
public List<flag> getScrollData(int start, int count) {
    List<flag> lisTb_flags = new ArrayList<flag>();      // 创建集合对象
    db = helper.getWritableDatabase();                    // 初始化 SQLiteDatabase 对象
    // 获取所有便签信息
    Cursor cursor = db.rawQuery("select * from tb_flag limit ?,?",
        new String[] { String.valueOf(start), String.valueOf(count) });
    while (cursor.moveToNext())                           // 遍历所有便签信息
    {
        // 将遍历到的便签信息添加到集合中
        lisTb_flags.add(new flag(cursor.getInt(cursor
            .getColumnIndex("_id")), cursor.getString(cursor
            .getColumnIndex("flag"))));
    }
    return lisTb_flags;// 返回集合
}
/**
 * 获取总记录数
 *
 * @return
 */
public long getCount() {
    db = helper.getWritableDatabase();                    // 初始化 SQLiteDatabase 对象
    Cursor cursor = db.rawQuery("select count(_id) from tb_flag", null);
                                                          // 获取便签信息的记录数
    if (cursor.moveToNext())                              // 判断 Cursor 类中是否有数据
    {
        return cursor.getLong(0);                         // 返回总记录数
    }
    return 0;                                             // 如果没有数据，则返回 0
}
/**
 * 获取便签最大编号
 *
 * @return
 */
public int getMaxId() {
```

```
            db = helper.getWritableDatabase();      // 初始化 SQLiteDatabase 对象
            Cursor cursor = db.rawQuery("select max(_id) from tb_flag", null);
                                                    // 获取便签信息表中的最大编号
            while (cursor.moveToLast()) {           // 访问 Cursor 类中的最后一条数据
                return cursor.getInt(0);            // 获取访问到的数据,即最大编号
            }
            return 0;                               // 如果没有数据,则返回 0
        }
    }
```

- flag.java 文件代码如下:

```
package com.wenlong.DBLab.model;

public class flag {
    private int _id;                                // 存储便签编号
    private String flag;                            // 存储便签信息

    public flag()                                   // 默认构造函数
    {
        super();
    }
    // 定义有参数的构造函数,用来初始化便签信息实体类中的各个字段
    public flag(int id, String flag) {
        super();
        this._id = id;                              // 为便签编号赋值
        this.flag = flag;                           // 为便签信息赋值
    }
    public int getid()                              // 设置便签编号的可读属性
    {
        return _id;
    }
    public void setid(int id)                       // 设置便签编号的可写属性
    {
        this._id = id;
    }
    public String getFlag()                         // 设置便签信息的可读属性
    {
        return flag;
    }
    public void setFlag(String flag)                // 设置便签信息的可写属性
    {
        this.flag = flag;
    }
}
```

- AndroidManifest.xml 配置文件代码如下:

```
<?xml version="1.0" encoding="utf-8"?>
```

```xml
<manifest xmlns:android="http://schemas.android.com/apk/res/android"
    package="com.wenlong.DBLab.activity"
    android:versionCode="1"
    android:versionName="1.0" >

    <uses-sdk
        android:minSdkVersion="8"
        android:targetSdkVersion="18" />

    <application
        android:allowBackup="true"
        android:icon="@drawable/ic_launcher"
        android:label="@string/app_name"
        android:theme="@style/AppTheme" >
        <activity
            android:name="com.wenlong.DBLab.activity.MainActivity"
            android:label="@string/app_name" >
            <intent-filter>
                <action android:name="android.intent.action.MAIN" />

                <category android:name="android.intent.category.LAUNCHER" />
            </intent-filter>
        </activity>
        <activity
            android:label="便签信息"
            android:icon="@drawable/ic_launcher"
            android:name="com.wenlong.DBLab.activity.ShowInfo">
        </activity>
        <activity
            android:label="添加便签"
            android:icon="@drawable/ic_launcher"
            android:name="com.wenlong.DBLab.activity.InsertFlag">
        </activity>
        <activity
            android:label="便签管理"
            android:icon="@drawable/ic_launcher"
            android:name="com.wenlong.DBLab.activity.ManageFlag">
        </activity>
    </application>

</manifest>
```

7.4 实验报告

1. 每人一份实验报告，统一使用学校提供的 A4 尺寸的实验报告册书写或者使用 A4 尺寸的纸打印。如果需要打印，则必须采用如表 E7.1 所示的表头格式。

表 E7.1 表头格式

实验课程						
实验名称						
实验时间	学年	学期	周 星期		第 节	
学生姓名		学号			班级	
同组姓名		学号			班级	
实验地点		设备号			指导教师	

2．实验内容的主要结果及对结果的分析。
3．实验过程中所遇到的问题的解决办法。
4．心得体会、意见和建议。

7.5　实验成绩考核

1．考勤方面的得分占总分的 10%。
2．相互协作完成实验任务方面的得分占总分的 40%。
3．实验报告方面的得分占总分的 50%。

附录 A　"智能终端应用程序开发"在线金课

在"智能终端应用程序开发"的教学过程中采用在线金课的形式将打破传统教学模式中时间紧、内容多、编程难等特点，彻底实现手把手教编程、开源代码丰富、合作开发项目、扩展软件开发能力等优势。"智能终端应用程序开发"在线金课与本书完全融合，以"滚动在线课程视频、在线试题库、在线案例源代码分享、在线讨论小组和共享提升资源库"的五位一体新平台模式推进教学活动。

"智能终端应用程序开发"在线金课使学生在较短的时间内进行 Android 开发环境的搭建，深刻理解 Android 平台体系结构，熟练使用 Android 基本组件、Android 的存储操作、多媒体开发、网络应用程序开发等技术进行在线学习、反复练习，并稳固掌握。

A.1　课程访问方式

（1）计算机端访问：可以登录"学银在线"网站，地址为http://www.xueyinonline.com/，在注册/登录后，在搜索框内输入"智能终端应用程序开发"关键字来查询课程，如图 A.1 和图 A.2 所示，然后单击图 A.3 中的"加入课程"按钮，申请加入课程。

图 A.1　学银在线首页

图 A.2　课程搜索结果

（2）手机端访问：可以下载超星学习通来进行移动端课程学习。目前，超星学习通支持 Android 和 iOS 两大移动操作系统。在下载并安装超星学习通之前，请确定您的设备符合系统要求。可以通过以下 3 种方式下载超星学习通。

图 A.3　"智能终端应用程序开发"课程首页

① 使用浏览器访问链接http://app.chaoxing.com/。该地址会永久存在，且内容会随着版本的更新而同步更新。

② 在手机的"应用商店"中搜索"学习通"。

③ 扫描二维码，跳转到对应链接后下载 App 并安装（如果使用微信扫描二维码，请选择"在浏览器打开"命令），如图 A.4 所示。

图 A.4　学习通下载二维码

A.2　在线金课课程体系

"智能终端应用程序开发"在线金课课程体系由在线资料、在线作业、在线考试、在线讨论、在线直播、公告通知、在线活动和课程统计共八大模块组成。

（1）在线资料：在线资料又分为 4 部分，即在线视频课程、课程介绍（课程大纲、课程宣传片、考试大纲、评分标准）、在线电子书资源、课程的 PowerPoint 文件下载。在线视频课程截图如图 A.5 所示。

图 A.5　在线视频课程截图

(2) 在线作业：教师在线作答并在线批改。部分在线作业截图如图 A.6 所示。

图 A.6　部分在线作业截图

(3) 在线考试：配有考试题库，要求学生在规定时间内进行在线考试。

(4) 在线讨论：教师可以分享试题并与学生进行讨论，同时学生也可以发布话题、共享知识。课程在线资源与话题讨论截图如图 A.7 所示。

图 A.7　课程在线资源与话题讨论截图

(5) 在线直播：教师在线直播并与选课学生进行在线互动，如图 A.8 所示。

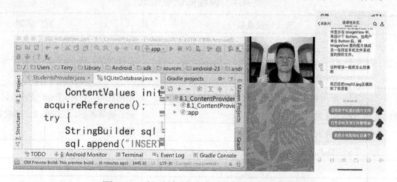

图 A.8　部分在线直播与在线互动截图

(6) 公告通知：通知每一次章节相关作业、测验和期中/末考试等相关信息。

(7) 在线活动：包括签到、投票/问卷、抢答、选人、评分、活动库等。在线活动界面截图如图 A.9 所示。

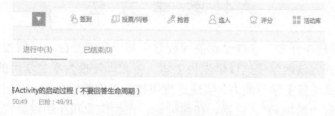

图 A.9　在线活动界面截图

（8）**课程统计**：包括已发布任务点、章节学习次数、章节测验、学生管理、讨论、成绩管理、教学预警、学习通活动、学习通积分等。课程统计管理界面截图如图 A.10 所示。

图 A.10　课程统计管理界面截图

A.3　教　学　内　容

"智能终端应用程序开发"在线金课分为理论、实验两部分，共计 11 章内容，包含 44 个在线视频知识点（462.3 分钟）。将用于授课的 PowerPoint 文件进行拆分与简化，以保证每个知识点对应的视频长度为 10 分钟左右，并在录制视频中突出最重要的文字、图形和动画。课程知识点概况如表 A.1 所示。

表 A.1　课程知识点概况

序号	章	节（知识点单元）	对应资源
1	第 1 章前言	1.1　前言	在线视频、PowerPoint 文件、阅读资料
2		1.2　什么是安卓	
3		1.3　课程特别说明	
4	第 2 章软件原型开发	2.1　应用程序设计与开发的基本理念	在线视频、PowerPoint 文件、讨论、自测题、作业、教学辅导、在线作业、在线答疑、在线测试、案例
5		2.2　"意大利塔"实验	
6		2.3　"意大利塔"实验的启示	
7		2.4　Balsamiq Mockups	
……	……	……	……
41	第 11 章实验	11.5　实验五：图形图像与多媒体	在线视频、实验指导书、项目任务、讨论
42		11.6　实验六：Android 的网络编程基础	在线视频、实验指导书、阅读资料
43		11.7　实验七：SQLite 和 SQLiteDatabase 的使用	在线视频、实验指导书、测试
44		11.8　实验八：使用 GPS 与百度地图	在线视频、实验指导书、百度开发者平台

A.4　教　学　方　法

"智能终端应用程序开发"在线金课的课程教学基于在线平台，采用校内课堂翻转、校外在线翻转的教学模式。该教学模式以学生为主体，将教师转换为学生学习的设计者和辅助者。该教学模式本着"提高自主学习能力、实现进度可控、重构教学环节、及时反馈教学问题"的原则，通过在线课程"微视频"、直播、在线讨论、在线作业和在线测验实现知识传授，将课堂翻转并以"任务驱动"实现知识的内化，从而达到预期的教学效果。这种教学方法主要分为课前传授知识、翻转巩固知识、课后拓展知识三大模块，如图 A.11 所示。

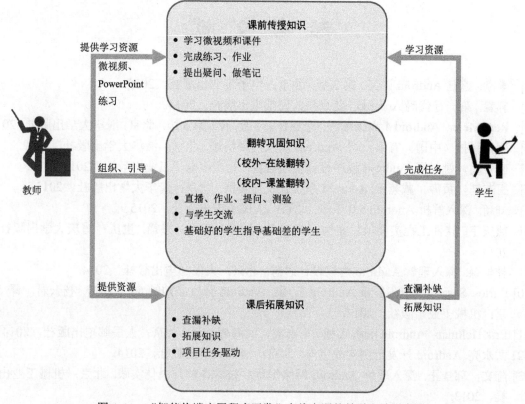

图 A.11 "智能终端应用程序开发"在线金课的教学方法示意图

参 考 文 献

[1] 李刚. 疯狂 Android 讲义. 第 3 版. 北京：电子工业出版社，2015.
[2] 郭霖. 第一行代码 Android. 北京：人民邮电出版社，2014.
[3] Reto Meier. Android 4 高级编程. 佘建伟，赵凯，译. 第 3 版. 北京：清华大学出版社，2013.
[4] 明日科技（中国）有限公司. Android 从入门到精通. 北京：清华大学出版社，2012.
[5] 向守超，姚骏屏. Android 程序设计实用教程. 北京：电子工业出版社，2013.
[6] 王友钊，黄静，戴燕云. Android 系统开发与实践. 北京：清华大学出版社，2013.
[7] 刘超. 深入解析 Android 5.0 系统. 北京：人民邮电出版社，2015.
[8] 熊庆宇. 软件工程实训项目案例——Android 移动应用开发篇. 重庆：重庆大学出版社，2014.
[9] 林学森. 深入理解 Android 内核设计思想. 北京：人民邮电出版社，2014.
[10] Carlos Sessa. 打造高质量 Android 应用：Android 开发必知的 50 个诀窍. 杨云君，译. 北京：机械工业出版社，2014.
[11] Erik Hellman. Android 编程实战. 丁志虎，武海峰，译. 北京：人民邮电出版社，2014.
[12] 武永亮. Android 开发范例实战宝典. 北京：清华大学出版社，2014.
[13] 陈文，郭依正. 深入理解 Android 网络编程：技术详解与最佳实践. 北京：机械工业出版社，2013.
[14] 郭金尚. Android 经典项目案例开发实战宝典. 北京：清华大学出版社，2013.
[15] Satya Komatineni，Dave MacLean. 精通 Android. 曾少宁，杨越，译. 北京：民邮电出版社，2011.
[16] 佘志龙. Google Android SDK 开发范例大全. 北京：人民邮电出版社，2011.
[17] 邓凡平. 深入理解 Android：Wi-Fi、NFC 和 GPS 卷. 北京：机械工业出版社，2014.
[18] 李钦. 基于工作项目的 Android 高级开发实战. 北京：电子工业出版社，2015.
[19] 姚尚朗，靳岩. Android 开发入门与实战. 第 2 版. 北京：人民邮电出版社，2013.
[20] 软件开发技术联盟. 软件开发实战：Android 开发实战. 北京：清华大学出版社，2013.